Beyond structure
The Power and Limitations of Mathematical Thougt in Common Sense,
Science and Philosophy

Europäische Hochschulschriften
Publications Universitaires Européennes
European University Studies

Reihe XX
Philosophie

Série XX   Series XX
Philosophie
Philosophy

Bd./Vol. 449

PETER LANG
Frankfurt am Main · Berlin · Bern · New York · Paris · Wien

Louk Eduard Fleischhacker

Beyond structure

The Power and Limitations of Mathematical Thougt in Common Sense, Science and Philosophy

PETER LANG
Europäischer Verlag der Wissenschaften

Die Deutsche Bibliothek - CIP-Einheitsaufnahme

Fleischhacker, Louis Eduard:
Beyond structure : the power and limitations of mathematical thought in common sense, science and philosophy / Louis Eduard Fleischhacker. - Frankfurt am Main ; Berlin ; Bern ; New York ; Paris ; Wien : Lang, 1995
  (European university studies : Ser. 20, Philosophy ; Vol. 449)
  ISBN 3-631-47900-X
NE: Europäische Hochschulschriften / 20

ISSN 0721-3417
ISBN 3-631-47900-X
© Peter Lang GmbH
Europäischer Verlag der Wissenschaften
Frankfurt am Main 1995
Alle Rechte vorbehalten.

Das Werk einschließlich aller seiner Teile ist urheberrechtlich geschützt. Jede Verwertung außerhalb der engen Grenzen des Urheberrechtsgesetzes ist ohne Zustimmung des Verlages unzulässig und strafbar. Das gilt insbesondere für Vervielfältigungen, Übersetzungen, Mikroverfilmungen und die Einspeicherung und Verarbeitung in elektronischen Systemen.

TABLE OF CONTENTS

Preface	7
Introduction	11
1 Mathematics as metaphysics	11
2 The problem of mathematical thought	12
3 The age of mathematism	14
4 What is mathematism?	15
5 The 'deconstruction' of mathematism	17

Chapter 1
What is mathematical thinking? 19
   1.1 Mathematical objectivity and mathematical abstraction 19
   1.2 Mathematics and mathematical thinking 27
   1.3 Philosophies of mathematics and of mathematical thinking 36
   1.4 An attempt towards synthesis 43
   1.5 Mathematical thinking and logic 46
   1.6 Mathematical thinking and technology 51
   1.7 The limits of mathematical thinking 56

Chapter 2
Degrees of reflection: statics 59
   2.1 Language, intelligibility and conceptuality 59
   2.2 The traditional doctrine of the degrees of abstraction 62
   2.3 Abstraction and reflection 65
   2.4 Empirical reflection 69
   2.4 Mathematical reflection 76
   2.5 Philosophical reflection 80

Chapter 3
The degrees of reflection: examples of the dynamics 91
   3.1 Between empirical and mathematical reflection 91
   3.2 Mathematical views of experience and philosophy 95
   3.3 Mathematism in modern philosophy 105

Chapter 4
The degrees of reflection: systematics 117
   4.1 Levels, dimensions and moments 117
   4.2 Philosophy and experience 121
   4.3 Anthropological considerations 123

Chapter 4
The degrees of reflection: systematics   117
   4.1 Levels, dimensions and moments   117
   4.2 Philosophy and experience   121
   4.3 Anthropological considerations   123
   4.4 The philosophical point of view   125

Chapter 5
Overcoming the mathematical paradigm in metaphysics   137
   5.1 Metaphysics in a mathematical style, and its fate   138
   5.2 The influence of the mathematical paradigm on metaphysics   142
   5.3 Against metaphysics   147
   5.4 Learning the lesson   156
   5.5 The principle of structurability   160
   5.6 The metaphysical attitude   163
   5.7 Metaphysics as the search for the principle of principles   166
   5.8 Principles and intersubjectivity   177

Index of persons   183

Index of subjects   187

Literature   211

Preface

My aim in writing the present work has been to give concise expression to the result of some twenty years of thought regarding the relationship between mathematics and metaphysics. I have always been fascinated by both disciplines and have found it difficult to accept that they were so compatible in Plato's time while apparently being rivals now. In this book I shall try to explain why this is so, and what influence it has on the world view of today. Logically, the next move would be to try to reconcile mathematics and metaphysics once again. Before that can be done, however, many obstacles must be removed. Most of them are of a conceptual nature; others have to do with current opinions, according to which the project is hopeless to begin with. I hope to be able to make it clear that such opinions are based on the very world view that is connected with the apparent animosity between mathematical and metaphysical thought in our time. In other words, most of the modern and post-modern arguments against metaphysics are based on the implicit conviction that mathematics is the adequate metaphysics. Even if metaphysics is rejected in the name of empirical science, it is usually the tendency towards a *mathematical* interpretation of experience which is responsible for this rejection.

This book has been written primarily for professional philosophers, though it should also interest those who are genuinely inspired by philosophy in their scientific or scholarly work, but have not specialised in it. Yet some elementary knowledge of mathematics and its history will be indispensable for understanding this book.

A little knowledge is a dangerous thing. But nowadays in science, not to mention philosophy, we seem to be condemned to it. We may have mastered one discipline, but that discipline borders upon innumerable others, of which we know almost nothing. And even if one dives deeper into ones own discipline, the number of questions that arise increases faster than the number of answers.

The central line of thought in this book is based on the very old idea that there are three principal levels, dimensions or degrees of knowledge: the empirical, the mathematical and the metaphysical. In Chapter 2 I shall try to summarize some of the historical shapes this idea has taken, though historical investigation is not the aim of this work. The aim of the present work is, rather, to disentangle contemporary confusions over the status of human knowledge knowledge.

This is by no means a purely academic enterprise. Many real conflicts between universes of knowledge are fought with epistemological weapons, in the attempt to undermine the status of the enemy's knowledge. Consider the conflict between the 'technocrats', who claim that there are 'scientific' solutions to many of society's problems, and those who oppose them with 'ethical', 'social', 'ecological' or 'cultural' arguments. The traditional conflict between religion and Enlightment is still raging since the eighteenth century, taking different shapes in the course of time. Governments see economic growth as a social necessity in order to create well-being; some philosophers on the other hand argue that this process creates its own scarcity and unhappiness. Psychiatrists of a Freudian bent regard personal frustration by social norms as the cause of violence, whereas those of a more structuralist conviction interpret the norms themselves as frozen violence, and explain the increase of patent violence in terms of their being loosened. In psychology, statistical data and phenomenological descriptions cross swords; and in the foundations of physics holists and positivists keep talking at cross purposes, never guessing their common presuppositions. The debate between the 'cognitive sciences' and philosophical anthropology concerning the meaning of 'artificial intelligence' socially reflects itself in two attitudes towards automation: should it replace or complement human resources?

All these debates are inconclusive, because the participants switch between different conceptions of knowledge, apparently disagreeing about some meaningful question, but really entangling themselves more and more in an intricate network of extreme positions. It is a task of philosophy to disentangle this network, and it is with the help of the concept of the traditional three 'degrees of abstraction' that I shall try to contribute to this task. Although the tool is a classical one, its use here will be very unconventional. In the first place, it has to be explained in such a way that its distinctions are recognizable for a contemporary reader, not only in theory, but in the actual practice of following discussions on such issues as those mentioned above. This requires systematization as well as illustration in various fields. Since not all of these fields will be relevant for every reader, illustrations and additional explanations which are not likely to be of relevance to all readers will be printed in a smaller type, and they may be skipped.

Each of the three levels of knowledge distinguished here entail a style of thinking which affects the perspective in which *everything* is seen, including the very distinction of the levels themselves. One might say that expressions such as 'levels', 'kinds' and even 'perspectives' all belong to the thought-style of the first level, the level of empirical concepts. Perhaps the expression 'dimensions' is characteristic of the second, mathematical, level; and the term 'moment' in Hegel's sense is an attempt to shape a language for the third. This shift of terminology is also a sign of a decreasing degree of *separability* of the forms of knowledge. This constitutes a well known philosophical problem, already indicated in Plato's cave-analogy. If philosophical knowledge understands itself as capable of attaining the highest degree of truth, how can it recognise other forms of knowledge at its side? Or should we take the inseparability of the forms of knowledge more seriously, including their inseparability from practical life? Perhaps, but I am not prepared to give up philosophy in favour of some conception of 'practice'. For however one interprets, say, Wittgenstein's thoughts about this point, they would have been inconsistent if he had meant 'practice' in such a way as to *exclude* something. Philosophising is clearly a human practice, and why should we want to be 'cured' of it? What hidden ethics distinguishes meaningful and nonsensical practices?

In this book we shall deal with all these conceptual difficulties, and with many more besides. And perhaps this is the main use to which this book may be put: to serve as a training ground for *systematic* thinking, without leading to a closed or rigorous *system*. Systematic thinking is here understood as the endeavour to find coherence in the conceptual framework used, without trying to construct such coherence beforehand. The coherence should be a sign of the unity of being, not of our urge for unification.

The present work is purely systematical, though it is of course impossible to philosophise in a systematic way without considering history, just as it is impossible to investigate the history of thought without systematic instruments. But the thinkers of the past are not quoted in this book for their own sake, but only for the actual value of their thoughts in the context of a systematic problem. Thus their role is somewhat like that of the 'authorities' of the Middle Ages, whose names are used as *identifiers* for certain clusters of current conceptions. So, when I quote Plato in order to illustrate the possibility of amazement about the status of mathematical objectivity, I do not claim historical knowledge about the actual role of this amazement in his thought. It is *my own* amazement, which I find adequately expressed in Plato's text. Of course there are limits to reading what one wants to read in philosophers' texts. But these limits are not determined by historical knowledge alone. They also depend on the systematic intricacies of the problems with which one deals. Of course one should avoid drawing historical conclusions on systematic grounds alone. On the other hand, one should also avoid innocence with respect to systematic questions in historical investigations. If the systematic context is coarse, interpretations tend to be equally coarse; if questioning is refined, interpretations become more refined, which is necessary but not sufficient for their historical correctness. For a systematic enterprise, however, striving for historical correctness is useful precisely in so far as it provides systematic inspiration. And it may do so, because a historically correct interpretation is more likely to widen our systematic horizon than a projection of our own ideas onto a text from the past. For the historical enterprise it is the other way round. If we are systematically naive, we run more risk of projecting our conceptions onto history. This precisely distinguishes the two disciplines. Only by observing this distinction they can be fruitful for each other.

I shall restrict the words of thanks, which tend to sound almost obligatory in prefaces, to three persons who contributed most to the coming into being of this book. Albrecht Kwast, who tirelessly encouraged me to complete it; Craig Dilworth, who commented acutely on the text and saved me from a host of linguistic blunders; and last but not least professor Jan Hollak, who inspired my philosophical thoughts for many years.

Introduction

1 Mathematics as metaphysics

It is sometimes claimed as an advantage, and sometimes regretted, that modern natural science has no metaphysics. In this book the unconventional thesis will be defended that *mathematics* has effectively functioned as the metaphysical foundation of the modern scientific tradition. The still living fundamental principle[1] of science, from Galileo onward, is the reduction of qualitative phenomena to measurable quantities and structures. Many underlying forms of thought in which this principle has been active apart from actual mathematisation, such as the mechanistic view, determinism, and positivism, have been superseded by others such as complementarity, probabilism, chaos-theory, critical rationalism and even sociologism. But the idea that knowledge is scientific in the complete sense of the word only if it is expressible in mathematical structures and equations seems to be unchallenged. Seemingly extreme reactions to the mathematical perspective,[2] such as holism, implicitly presuppose the same principle of measurability, which precisely accounts for their apparent extremeness. Even if real mathematisation lies far behind the horizon, as it does in the cognitive sciences, it is nevertheless taken as a standard, e.g. in the form of computational models. In logic and linguistics as well - and even in ethics - the mathematical perspective is prevailing now in the form of what is called 'formalisation' or 'formal methods'.[3]

---

1 The term 'principle' is of central importance in this book. Its use will be explained extensively in Chapter 2. In a preliminary way it can be clarified by saying that a principle is the origin of a *perspective* in which the whole of human experience can be interpreted. Such a perspective does not *exclude* factual possibilities as a scientific theory does. The example of the mathematical perspective, originating in what I will call the *principle of structurability*, will be elaborated extensively in chapter 1.

2 The notion of 'perspective' plays a central role in this book. It has been suggested to me as important in the first place by prof. H.M.J. Oldewelt (see note 51) in 1960. Later I learned that it was used by others too as a central notion. See for instance Craig Dilworth's *Scientific Progress*. A contribution to 'perspectivism' which I claim as my own is the idea that perspectives originate from the implicit insight into principles on the one hand, and that principles are *relatively* transcendental with respect to those perspectives on the other. See Chapter 5 for the extensive explanation of this idea.

3 For a coherent and convincing criticism of this trend, see: Sören Stenlund, *Language and Philosophical Problems*, 1990.

2 The problem of mathematical thought

From the perspective of the philosophy of mathematical thought, however, the relationship of mathematical structure to observable reality has remained extremely problematic. Plato formulated the question where in the world to look for numbers and geometrical figures,[4] and he concluded that the visible world is not the only possible mode of being. Mathematics cannot be about the world of human experience, for example, because this world resists reduction to purely mathematical structure. The reality of change is an especially hard nut to crack, as was noticed already by Aristotle. But he found a way different from Plato's for dealing with mathematical objects. He regarded them as the results of abstraction, the actualization *in thought* of a principle we find in the world of experience. He called this principle 'ὐλη νοητη - intelligible matter - and it will be discussed in Chapter 1.

In antiquity the main problem was to *separate* mathematical objectivity from experience, without making the applicability of mathematics impossible to understand. Modern times, however, begin with the idea of the *identity* of mathematics and physics. Nature herself is thought to be structural, and thus accessible to mathematical investigation, not only by her external geometrical shapes - as Archimedes had already discovered - but also in her inner laws.

In philosophy this had a very strong impact. Descartes characterises the world of experience as *res extensa*, taking what in Aristotelianism had only been an outer property of material things to be their essence. The externality of nature becomes its inner principle. Hegel in the nineteenth century formulates the essence of nature as 'the Idea in the form of externality to itself'.

For philosophy this meant that the problem was no longer one of the relation of the mathematical to the physical, but of the knowing subject, Descartes' *res cogitans*, to a mathematical=physical world. This produced a strongly mathematically coloured, but never really mathematical, metaphysics.

Spinoza's attempt to construct a metaphysics *'more geometrico'* has led to points of view which actually went beyond mathematical reasoning, but remained strongly influenced by it. Even when in modern philosophy the paradigm of geometry, or mathematics in general, is explicitly rejected, as in the case of Hegel's system, the lure of structural rigour is still present.

In the nineteenth century the identification of mathematical and physical objectivity became less and less obvious. Mathematics, liberated from its close connection to physics and technology, began to develop highly speculative theories such as complex number theory, abstract algebra, Fourier analysis, non-Euclidean geometry and projective geometry. It started looking for a

---

4   Plato, Republic 526a

foundation of its own, independently of physics, and thereby more and more overtly showed its ideal character.

In philosophy, on the eve of the twentieth century, two - apparently opposite - impulses emerge, which may eventually undermine the ideal of mathematical rigour: Husserl's phenomenology, which introduces another ideal of philosophical rigour, and Frege's mathematical logic, which *objectifies* mathematical reasoning.

Gödel's results teaches us that, as a consequence of this objectification,[5] the foundation of mathematics cannot be formulated explicitly as a mathematical theory. Mathematical thought as such cannot be free from intuitive presuppositions demanding investigation by a discipline other than mathematics itself.

For a 'working mathematician' this is no problem at all. She or he is perfectly happy with Hilary Putnam's *'Yes, we have no foundations'*; but the philosopher experiences again a change of problem-field. Now *both* the subject-object relationship *and* the relationship of structure and reality have become problematic. The mathematical point of view appears to be based on an intuitive insight, constituting a certain perspective - which I call *structural*, for reasons to be explained later - on the world of experience.

But if that is true, mathematical structure is not necessarily the only or even the most adequate form in which scientific knowledge can be expressed. Perhaps the success of measuring-science has blinded us to metaphysical perspectives, whether or not they justify or radicalise the mathematical approach. Even philosophies which are generally considered to be anti-mathematical, such as Hegel's speculative dialectics or Heidegger's existential philosophy, when inspected more closely, appear to share certain essential presuppositions with the mathematical approach, e.g. the denial of real potentialities. In fact the ideal of 'exactitude' - the possibility of making all presuppositions explicit and developing a body of thought consistently from them - seems to be all-pervading in our culture. Wittgenstein's *Philosophical*

---

5 The words 'subject' and 'object' are used in different senses, but the tendency always is that they are correlatives in the performance of some (theoretical or practical) action - as the linguistic use suggests. The subject is the active pole, the object not necessarily passive, but the action is always directed towards it. Subjective is what belongs to the subject as such (i.e. in its function of being the active pole), objective what belongs to the object as such (i.e. in its function of being the 'aim' of the activity), which does not necessarily mean that it exists independently of the subject. Mathematical objects for instance, need not be conceived of as existing independently of mathematical thought. Subjectivity and objectivity are the properties of being subjective, respectively objective. Objectification is the act of giving objectivity to some content, either by theory - conceiving of it in the form of objectivity - or by practice - bringing about a state of affairs which may be understood as representing the said content in an objective form.

Investigations, because of its anti-systematic tendency, may be regarded as an outstanding exception; but this work also shows clearly the kind of trouble that arises if one tries to leave the mathematical paradigm behind. For what other method than allusion remains, if an explicit development of ideas is forbidden? The very different ways in which Wittgenstein's philosophy has been interpreted make this clear, for if one cannot express one's ideas explicitly, there is no limit to interpretation.

Also in structuralism, in spite of its name, a tendency to leave mathematical grounds is present. It is *real* structure the structuralists are after, not ideal, mathematical structure. But as long as *nothing but* structure is seen, it is already surreptitiously being idealised. Therefore, in structuralism there is always an essential non-structural principle - such as power, force or spontaneity - lurking in the background. Critics of the the mathematical point of view usually underestimate its power. Either it eventually turns out that they have remained within it or they adopt its abstract opposite and in this way remain indebted to it.

3 The age of mathematism

As the whole enterprise of this book is meant to be purely *systematic*, this introduction is the proper place for historical remarks. This is especially true as regards the claim that the modern era - which as a non-historian I allow myself to take as being from 1500 to the present - is *not* the era of the mechanistic world view, *nor* the era of materialism, but the era of mathematism. Dijksterhuis concludes his *Mechanisation of the World Picture* with the remark:

> The mechanisation, which the world-picture underwent in the transition from antique to modern natural science, consisted in the introduction of a description of nature by means of the mathematical concepts of classical mechanics; it indicates the beginning of the mathematisation of natural science, which obtains its completion in twentieth century physics.

But this is only seen from the direction of the ultimate *effect*. In my view, the technical as well as the philosophical sources of the rise of modern science already introduced very strong tendencies towards mathematisation. As we shall see in Chapter 1, it is in accordance with the natural development of technology that technical concepts are made more and more explicit. Of course this does not explain the particular historical period in which this development took place. But one thing is clear: in order to make technical concepts explicit, one must measure and calculate. Moreover, on the philosophical side, medieval Aristotelianism hardly left room[6] for another basis to be criticised on than precisely the mathematical Platonism that arose in the Renaissance. The breakthrough of both tendencies - the technological development and mathematical Platonism - and their fruitful meeting in a particular place and time can probably be explained by fundamental changes in society.[7] What is important here, however, is the result of the breakthrough: the firm belief that measurement and mathematical calculation, and nothing else, will lead to insight into the phenomena of nature. For Galileo the book of nature was written in mathematical signs, and for Newton mathematical space and time were absolute, whereas experienced space and time were considered to be only relative. Nature came to be seen as mathematical *in itself*, and the distinction between mathematics and physics became obsolete. In the eighteenth century 'mathematics' still encompassed a whole range of disciplines, from arithmetic to machine-construction. Only in the nineteenth century did a new form of 'pure' mathematics emancipate itself from natural science and technology. But by then the mathematical style of thinking had been thoroughly spread among scientists and technicians.

4 What is mathematism?

But the prevalence of a particular style does not itself constitute mathematism, which is rather a - usually implicit - metaphysical position connected to the feeling that the 'mathematical' style is so self-evident that it does not need any foundation. In this way this style is itself taken as the foundation of science. As a consequence, the objectivity and generality of the style have to be regarded as objectivity and generality without qualification. As I shall explain later, the object of mathematical thought can be characterised as *structure*, which is more general than what is usually understood by quantity, but is by no means identical to metaphysical universality. If unqualified objectivity is identified with mathematical objectivity, the fundamental nature of reality becomes structure, which is differentiated only by higher or lower degrees of complexity. This is exactly in line with the philosophy, ascribed to Pythagoras, according to which the essence of the universe is *number*.

---

6 Especially for those who - like Cusanus and the humanists, and unlike most of the modern philosophers - knew perfectly well what it was all about, and where the strength and weakness of this world-picture was to be located.

7 Max Scheler, *Die Wissensformen und die Gesellschaft*

Number for the ancients was the principle of what is mathematical, and it is still often regarded as a fundamental paradigm of structure.[8] The Pythagorean world view is a fundamental and ever recurring metaphysical perspective. In Plato's Academy, Speusippos and Xenocrates took up this line of thought and in the Renaissance it was popular with humanists such as Pico della Mirandola. Even today it is explicitly adhered to by theoretical physicists, who doubt whether 'matter' is to be regarded as a useful concept in physics.

On the other hand we all know that structure is not something immediately given. We can see different structures in one and the the same phenomenon and we can technically give different structures to our surroundings. And in pure mathematics, structure is the result of postulation or thought-construction. So structure is in a certain sense our product. It is the *structurability* of the world, which is fundamental.

So mathematism has two sides to it, expressible in two ideal-typical theses:

1. Structurability is the *essence* of everything.
2. To know something is the same as to give it structure.

This is a completely coherent metaphysical position, in which being is identified with *mathematical* intelligibility, instead of intelligibility without qualifications. But the question is, whether or not this world view is unduly *restrictive*. Does it rule out any other perspectives which we find particularly plausible? One could ask whether anything exists which is not - in a certain sense - structurable, and it would be difficult to find an example. On the other hand, one could ask whether in fact there exists anything the essence of which *is* its structurability. Perhaps one could think that the essence of *space* is its structurability. But once one imagines something *in* it, a non-structural quality is introduced, which distinguishes the space occupied by the 'something' from empty space. Trying to reduce this quality to structure again, could very well lead to an infinite regress. If indeed it is felt as absurd from the point of view of common sense to express mathematism as an explicit philosophy - in the same way as it is felt as absurd to express scepticism[9] as an explicit position - what then are the grounds for this feeling of absurdity? It is just like the

---

8   In this connection Kronecker's saying: 'The natural numbers are made by the Lord, the rest [of mathematics] is human work' is usually quoted.

9   Scepticism disregards its own claim for truth. Therefore it is immediately refutable by showing its 'pragmatic' self-contradiction. But even then it is not refuted as a general *attitude* in life. Hegel saw this clearly in the chapter on scepticism in the *Phenomenology of Spirit*. Cf. also Michael N. Foster, *Hegel and Scepticism*, 1989.

aporia in debates about artificial intelligence. If one mentions a human skill, not yet simulable by computer programs, the AI defender will say: if you describe it exactly and clearly (i.e. mathematically) to me, I shall find a way to simulate it. But then, if it is so described, it is probably not the same as it was before. What, however, is the *difference*? We have the feeling that, as soon as we describe this difference, a corresponding correction of the program will eliminate it. We are so immersed in mathematism that we simply cannot imagine a kind of exactness *surpassing* mathematical exactness. For how could we prove that e.g. intelligence is *not* reconstructible in mathematical terms, if not by using a description of mathematical reconstructability itself, showing its restrictions. But such a description should evidently be clearer and more self-evident than any mathematical reconstruction. Only a level of thought traditionally ascribed to metaphysics could perform this task, and that is why mathematics and metaphysics are rivals in a mathematistic world. On the other hand, it seems to be precisely the development of information-technology that tends to change this situation. In this field structures are of course important, but they can no longer be considered as purely mathematical. They are not invented for the sake of clarifying the domain of the ideally structural, but for the sake of their use in a context of human practice. Mathematically, they are clumsy and opportunistic. They have nothing of the proverbial mathematical elegance, their adequacy cannot be rigorously proven and their functioning cannot be completely tested. Mathematicians, like metaphysicians, stand here awkwardly looking at something of which they claim to know the principles, but to which they cannot apply them. The two may become brothers again.

5 The 'deconstruction' of mathematism

The enterprise, of which this book is a report, consists of an attempt towards a systematic 'deconstruction'[10] of mathematism. The real philosophical problem one has to face in such an attempt is that 'deconstruction' is not such a neutral and unbiased activity as it is sometimes thought to be. It itself involves a metaphysical point of view, which cannot be left completely implicit. This point of view has to be distinguished from

---

10 This term is borrowed from Jacques Derrida. In his philosophy it indicates the enterprise of showing the context-dependent history of philosophical expressions. They are considered as the tracks of thought-constructions rather than the names of transcendental principles. To deconstruct a form of thought, therefore, is to show how it is constructed - or construed! As a basic philosophical attitude, deconstruction has to be rejected, as I hope to show in Chapter 5. But as a component of the critique of a specific philosophical position it can be very useful.

a number of traditional conceptions, in order not to be identified with one of them and thus be perceived as anachronistic. So this book is actually an attempt to make progress in metaphysics. It has become fashionable in our century to contribute to this field by rejecting the preceding tradition as a whole through the revealing of a fundamental mistake underlying all of it. My approach is the opposite. I shall try to find a common motive in a number of - pro- and contra-mathematical - metaphysical views, as the basis of an attempt to make a further step. What is common to all philosophy traditionally goes by the name of *philosophia perennis*. This is in our day one of the most unfashionable of expressions. But fashions change, and in any case I do not intend to be distracted from my original inspiration just because it happens to be out of fashion.

Chapter 1
What is mathematical thinking?

In this chapter a characterisation is given of mathematical thinking. We are not concerned with mathema*tism* here, but consider mathematical thinking as a legitimate way to obtain insight of a certain - important - kind. We shall try through philosophical reflection to get a clear picture of the nature of this kind of insight and of its connection with contiguous domains, such as technology, logic and artificial intelligence. The classical philosophical problem of the nature of mathematical objects is taken as a starting point.

1.1 Mathematical objectivity and mathematical abstraction

The characteristic of mathematical entities - as Plato already discovered[11] - is that they are at the same time ideal and individual. They are 'grins without a cat'[12] - ideal, though not thought of as ideal, but rather as realities in another world. It is difficult to avoid the mathematical idealism that grants the ideal its own sphere of reality. Even Aristotle's *deutera ousia* and Husserl's *ideales Sein* tend towards a separate existence, whatever is done to avoid this idea. This is due to the tendency to regard existence as univocal, which is quite natural for language and thought used to refer to the entities in the world around us. But if one makes the ideal too real, one loses it, as Orpheus lost Euridice. Wittgenstein described this using his metaphor of the beetle in the box:[13] if the meaning of a word is considered to be a (in this case psychical) separate entity, one can do without it. Whatever the meaning is, it can be neither 'outside' nor 'inside' the word. So it resembles a

---

11   Republic, 526a

12   "All right," said the Cat; and this time it vanished quite slowly, beginning with the end of the tail, and ending with the grin, which remained some time after the rest of it had gone.
"Well! I've often seen a cat without a grin," thought Alice; "but a grin without a cat! It's the most curious thing I ever saw in my life!" (Lewis Caroll, *Alice in Wonderland*, Chapter VI)

13   *Philosophische Untersuchungen* 293. Wittgenstein argues that if certain persons each possess a beetle, which they hold in a box and leave it there, the semantics of anything that can be said about those beetles never depends on their real existence. So if the meaning of a word were a purely individual psychic entity, is would be superfluous for its use.

geometrical body, which, according to Aristotle's famous proof,[14] can neither exist inside nor outside a physical body. It is an alien soul, which is physically embodied as the soul of the dead king in a pyramid.[15] Indeed, embodiment seems to be an adequate metaphor, but then the same problems arise again. 'Has our spiritual soul the same relation to our living body as a sailor to his ship?'[16] Aristotle already asks. In fact Plato also seems to have had this opinion, and there is evidence that he ascribed to the human soul an ontological status on the same level as mathematical objects.[17] This level lies between that of the material world and the ideas. The human soul and mathematical objects seem to connect these two, but the concept of 'level' is a mathematical metaphor too. Levels are separable from each other, not always in practice, but in principle. And that is precisely what mathematical objectivity is all about.

An apple is divisible into two halves. If we in fact divide it, the original apple as a whole no longer exists. But before we divide it, we *know* that it is divisible. And thus Aristotle describes *quantity*: 'That which is divisible into constituent parts, each or every one of which is by nature some one individual entity.'[18] Of course he is not thinking of any examples other than 'length,' 'surface,' 'volume' and 'number.' But the description is still adequate for what we nowadays call 'mathematical structure.' If one opens an arbitrarily chosen textbook on a mathematical discipline, one will probably find that this discipline is about certain domains of entities which are related to each other in certain ways. The relations have to fulfil certain rules or axioms, characterising the specific kind of structure dealt with in this discipline. The structures may be algebraic, topological, geometrical, etc.

But it is not only in abstract mathematics that we think of such structures. All analytical thought considers its subject matter as consisting of interacting parts, which can also be considered in their own right. In general the parts either do not have precisely the same properties within or without the whole (e.g. the constituents of a chemical compound) or cannot even be

---

14   Metaphysics XIII,2 [1076b-1077b]

15   G.W.F. Hegel, *Enzyklopädie der philosophischen Wissenschaften*, par. 458; Suhrkamp Werke Bnd. 10, p. 270, Cf. J. Derrida, Le puits et la pyramide, *Marges de la philosophie*, 1972.

16   De Anima II,1 [413a8]

17   Cf. Konrad Gaiser, *Platons ungeschriebene Lehre*, 1963.

18   Metaphysics, V,8 [1020a6]

meaningfully thought of under abstraction from the whole (we cannot e.g. consider properties without thinking of something *having* them)[19]. But often we find a way to eliminate these problems. We can think of the parts as themselves again consisting of parts, recombining in a different way within or without a larger whole. Or we can think about them as virtual entities, becoming only real in combination with other virtual entities.[20]

In this way it is often possible to explain the properties of whole by reference to its *structure*; and the next step might be to consider this structure in the abstract and try to apply mathematical reasoning to it.

If we call this process globally 'mathematical abstraction,' we may feel that a number of different operations are being rashly thrown in a heap and given one name. First we think of something as a whole consisting of parts;

---

19   In his commentary on Aristotle's description of the category of quantity (Metaphysics, V,8 [1020a6] Thomas Aquinas distinguishes the quantitative relationship of part and whole from 'mixtum' in which the parts influence each other in such a way that their properties within the whole are different from the properties they would have outside the whole, although the parts can actually exist without the whole. A chemical compound or an organism are clear examples of this. (We have to be aware of the fact that although sometimes a mathematical model of such situations can be conceived of, the model remains always a *reconstruction* of physical reality) Mixtum and quantitative compositum have in common that the parts can exist separately, but in the mixtum precisely in so far as they are *not* parts, and in the quantitative compositum in so far as they *are* parts: their existence *within* the whole is already an individual existence. On the other hand the quantitative compositum is distinguished from the *formal* composita, such as the hylemorphic compositum of matter and form in which the parts as such do not have an individuality of their own and cannot exist separately. (See Thomas Aquinas: In Metaph. Arist. 977 and Aristotle De Generatione et Corruptione 1034b20-1036a12) The concept of quantity or structure as a specific form of the relationship between part and whole is not characteristic for the Aristotelian tradition, although dealing with it in terms of the potency/actuality relationship is. Husserl, e.g. analyses in part II of the second book of his *Logische Untersuchungen* a great variety of part-whole relationships, among which the 'Zerstückung' in which the parts can be imagined as existing independently with respect to each other and to the whole, corresponds to the 'quantitative' part-whole relationship. Husserl recognizes in the last section of this part (par. 25, pp. 288-293) that *in physical* reality this form of part-whole relationship does not in fact exist. It only exists in the imagination, which according to my view is based on mathematical abstraction.

20   This is the way Frege understands the mathematical function-concept. He does not yet think of a function - as already many mathematicians do in this time - as *itself* a mathematical object. A function, according to him, is an *incomplete* entity, which requires completion by specification of the values for the free variables in its expression. This notion of incompleteness has later been transferred to the *symbol*, so that e.g. a verb can be regarded as an 'incomplete symbol', which needs completion e.g. by the subject and objects of the sentence. This gives rise to a mathematical linguistics in which the words are interpreted as function symbols.

then we rearrange things in such a way that the parts become to a certain extend *independent* of the whole; next we think of the new whole as of a structure; and finally we consider this structure as something *in itself*.

The first two steps, which may be characterised as 'structuring the world,' seem to be *directed towards* finding an ideal structure representing the phenomenon adequately.

Thus we may e.g. think of a process as being an accumulation of many partial processes. But such a division is arbitrary; and just because we *understand* in which respect it must be arbitrary, we are able to think of an ideal mathematical representation of the process by considering *all possible* divisions, i.e. by the infinitesimal calculus.

Another example is a sentence considered as divisible into words. By considering these words as not rigidly connected to their meanings, but as *transforming* each other's meanings according to certain laws, the meaning of the sentence may be explained as resulting from the structure consisting of a specific combination of these transformations.

So here already the notion of an ideal structure is implicitly present, just as the notion of number is present in every attempt towards measurement. It is even said[21] that every technical manipulation of material is connected with such a notion of ideal structure as a measure for the adequacy of the result or as itself a non-material result. In any case, our practices, even of the most trivial, everyday' sort, include an element of structuring of the world. Therefore they also include implicit or explicit *knowledge* of the *possibility* to do so, knowledge of the *structurability* of the world.

This knowledge is the source of the second two operations: thinking of a phenomenon as being the implementation of a certain structure, and conceiving of an *abstract* structure. From the point of view of structurability, everything is a realisation of its own structurability. Something *has* a certain structure because it has been structured thus either by nature, or by human practice, or by human understanding. But then the structure plays the role of a *concept* by which the thing has been formed or by which it is understood. By looking at structure this way, we may grasp the idea of an unrealised, ideal concept. But if an ideal concept is thought of as *structure*, it is in need of some 'ideal material' which is thought of as structured according to this structure-

---

21   Cf. E. Husserl, *Die Krisis der europäischen Wissenschaften und die tranzendentale Phänomenologie*, 1954, Zweiter Teil, par. 8, pp. 21ff. See also Hugo Dingler, *Das Experiment: sein Wesen und seine Geschichte*, 1928.

concept. This ideal material was called 'intelligible matter'[22] by Aristotle, because it is thought of as a material of which structurability is the *only* property. This is why it is said that mathematical thought abstracts from sensory experience and from change *as such*, i.e. from those characteristics of our experience which cannot be regarded as the result of structuring. Perhaps the physical causes of change or of sensory experience can be reconstructed in a theory in which they are regarded as determined by specific structures, but *they are not experienced that way*, and ideal structures do not change and are not experienced by the senses.

Mathematical abstraction is the mental act by which we consider structural properties to be the substantial subject matter of thought. It is a traditional question of the philosophy of mathematics how this is possible without error. An answer to this question must involve epistemological as well as metaphysical considerations. The relation between the *result* of mathematical abstraction and the physical entity or interaction-process it is performed on will be described now.

Mathematical abstraction does not pick up ready-made structures in the physical world in order to transfer them into the realm of 'ideal' mathematical objectivity. Physical space for example is neither 'Euclidean' nor 'non-Euclidean,' for the question of the validity of the parallel-postulate has no physical meaning.[23] Another, even simpler example is the notion of an open or closed interval. As in the physical world nothing corresponding to the mathematical notion of a point exists; the difference between an interval that contains its end-points and one that does not has no physical meaning. This distinction depends only on the *principle* of structurability, which is abstracted from the world of experience,[24] whereas the *realisation* of the principle is of a totally different nature in this world than it is in the world of mathematical objectivity. Material structure appears as a phase in an interaction process, and the precision of its determination decreases with its increasing generality in time and space. Structure is only a spin-off from physical interaction, and it is never completed in that process. It can, however, inspire our thought towards the construction of completed, ideal structures. But this always involves arbitrary decisions, such as the ones between open and closed intervals, Euclidean and non-Euclidean geometry, Cantorian and non-Cantorian set-theory.[25] These ideal structures cannot be considered

---

22 'ὑλη νοητη, Metaphysics VII,10 [1036a8-12]

23 Being Euclidean or non-Euclidean is a *global* property of space. As measurement only reveals *local* properties, the difference cannot be discerned by measuring. Of course physics as a science can use Euclidean or non-Euclidean *models* of space, but this is merely a matter of the mathematical interpretation of physical quantities.

24 The term 'world of experience' is used in this book in such a way that everything that may be *understood* - but not necessarily *believed* - to be the content of some description of human experience belongs to this world. So in a sense it is a 'phenomenal world', and we abstract from the question of the 'reality or non-reality' of any of its contents.

25 These decisions always transcend the problem field for which the particular realm of structures is conceived. In Euclidean and non-Euclidean geometry, *local* properties are practically the same, and the question of the validity of the continuum hypothesis has

as a by-product of something else, they must be thought of as having an independent existence and as being completely determined in themselves.[26] Structural change and indeterminateness therefore must be reconstructed continuously in science, using ever more complicated mathematical structures. This mathematical ideality is for philosophers a hard nut to crack, but all attempts to avoid it have failed so far. Intuitionistic constructivism has developed into a branch of 'quasi-classical' mathematics, and Wittgensteinian finitism is only made more absurd by the attempts to defend it.[27] In fact all philosophical positions denying mathematical ideality are based on a refusal to recognise the distinction between mathematical and physical structure. Pythagorean positions however, construing mathematical structure to be the very essence of physical reality as such, have exactly the same prerogative, for failure to distinguish between mathematical and physical structure leads either to the tendency to reduce mathematics to physics or to the tendency to reduce physics to mathematics.[28]

Once the idea of pure ideal structurability is understood, the whole unbounded domain of mathematical objectivity is open to investigation. Of course, historically, there was not a single and total breakthrough. Mathematics

---

practically no bearing on the present body of mathematical knowledge. Yet mathematics and its applications are in continuous development, and the irrelevant questions of today can become the crucial ones of tomorrow.

26 Although *formal* axiomatics cannot express this complete determinateness because of the Skolem-Löwenheim theorem and the possibility of non-standard interpretations, *intentionally* it is the aim of axiomatics to express it. The intentional aim of Peano's axioms, e.g., is to determine in an absolute way what the system of natural numbers really is. The failure to formalise this does not prove that the intentional aim is not in fact reached.

27 Cf. S.G. Shanker, 'Wittgenstein's Remarks on the Significance of Gödel's Theorem', 1988, pp. 155-256. In this article Wittgenstein's justified criticism on metamathematics *as a foundation of mathematics* is stretched to become a criticism of metamathematics as such. On the other hand no convincing alternative foundation of mathematical thought is offered.

28 The confusion is quite clear in discussions round the '*actio in distans*' problem and round the 'locality principle'. These discussions necessarily involve a notion of 'contiguity'. But this may be interpreted either *mathematically* as 'having a common border', or *physically* as 'being in immediate interaction'. Now any physical interaction occupies space, so in the physical interpretation every action is in a certain sense '*actio in distans*' by definition. On the other hand the mathematical notion of 'touch' in a common border has no physical meaning at all. So if this conception is taken seriously, only bodies occupying the same space may interact. This leads to an ether-theory, in which eventually all matter is identified with the all-pervading ether and in which the concept of interaction becomes empty. The 'locality' problem is more intricate. The locality principle states that no information can be transferred faster than light. But this involves a *geometrical* notion of distance, which is not necessarily adequate for physics. Perhaps not only the notions of simultaneity and contiguity, but also the notion of distance has to be revised from a naive geometrical to a more 'physical' notion.

first existed as a practical discipline of measuring and counting. Later, in ancient Greek culture, it was developed as a science of ideal structure, but structure was understood rigorously as a possible geometrical or arithmetical structure of something given in experience. Moreover the continuous and the discrete were safely separated in the traditional disciplines of geometry and arithmetic. In modern times the ideal was reunited with physical reality, which was thought of as intrinsically mathematical. The immediate connection with phenomena and the separation of the discrete and the continuous were somehow swept away by this development. Mathematical thought as it emerged from this historical process in the nineteenth century was no longer hampered by an immediate connection to experience, and it could deploy its abstract structures ad infinitum. And with this result, ever more intricate ways of structuring the world emerged, partly as direct applications of results of this mathematical development, and partly as an effect of the inspiration it provided.

If mathematical abstraction is based on the intuitive knowledge of the principle of structurability, this principle must necessarily be a real principle[29] of the experienced world. The experienced world as such is a relative entity, because it is defined in relation to a subject of experience. But the relativity of something does not make it or its principles less real.[30] The reader should also bear in mind that a principle, even if it is claimed to be real, does not

---

29 I distinguish a 'real' principle from a merely methodological principle. Mathematical abstraction might be considered as a *method*, regulated by the principle of structurability. But if this method is more than an administrative or linguistic expedient, i.e. if it really provides *knowledge* of its own, structurability cannot be something which is arbitrarily *postulated*. It must be *real*, in the sense that what we *think* of as structurable, must really *be* structurable. And this must not be a superficial property or else mathematical knowledge would be equally superficial. But mathematical thought justly claims that it provides a rather profound level of knowledge. So why should one, as many scientists do, believe in it firmly in practice and depreciate it in theory?

30 Whatever is *objective* is objective *for a subject*. As **Hegel** formulates it: nothing on earth exists which is not objective and subjective at the same time. What is real for the human mind is determined by a subject/object relationship. Here the question of intersubjective validity might be raised. It cannot be dealt with in the framework of this book, but my opinion is that philosophers such as **Wittgenstein** and **Scheler** are right here in supposing that the private domain *presupposes* the public. Philosophers of the *modern* tradition, from **Descartes** to **Hegel**, seem to have presupposed this silently. Hegel himself on the other hand has formulated it explicitly: *self-consciousness necessarily exists for another self-consciousness*. By using language to express his ideas, every philosopher admits this in practice, however strong his denial of it in may be theory.

exclude any possible fact. It is not something like a falsifiable theory or model, but the origin of a perspective in which *all* facts can be interpreted.

We shall explore the nature of the structurability principle to some extent. It is regarded as enabling us to understand something given in our experience as a whole built up from a plurality of interrelated individual parts. These parts need not be spatial. They can be distinguished by whatever property we may choose. In fact every division of the experienced world into things having properties *is* already a structuring. When we use a characteristic to distinguish an entity, it thereby *becomes* a property of the entity because we have necessarily also distinguished the characteristic from that which it characterises. Quality and quantity, property and structure, are born together. As a result of structuring, the experienced world contains certain 'substantial elements,' divided and related in a certain way. We may think here of the distinction of 'natural kinds' or of things within such a kind or of parts within a whole. Some philosophers object to what they call 'reification,' but structuring is always reification because it creates the distinction between 'entities' on the one hand and their properties and relations on the other.

In Aristotelian terms one could say that structurability causes the distinction of substance and accident. This does not imply that it determines *what* it is that exists as substance and *what* it is that exists as accident, nor that actual structuring *arbitrarily* makes this determination.[31]

A mathematical example is the structuring of experienced space. In Euclidean space lines and points are substantialised. In analytical geometry only points are substantialised, and lines are reduced to relationships between points. Yet both structures are essentially the same. Space may also be structured in a non-Euclidean way. This is no arbitrary alternative: both structures can be embedded in more abstract spaces, and even the most abstract

---

31 In his interesting but somewhat verbose book, *Two Logics*, Henry B. Veatch distinguishes the constructivistic tendency of scientific thought and its interpretation by philosophers of science from the realism of scholarship. His position is clarifying, but also exaggerated. Science, in his opinion, is based on what he calls a *relating logic*, which is adequate for dealing with relational constructions, whereas scholarship is based on a *what-logic*, which is able to express *what a thing is*. His exaggeration consists in his on the one hand detaching the relational constructions of science completely from the experienced world for the understanding of which they are conceived, and on the other, overlooking the constructive element in our common sense division of the experienced world with the help of categories such as Aristotle's or Kant's. Even if the categories in themselves are necessary, the subsumption of the objects of experience under them is not. The underlying problem is that it is difficult to concentrate thought on what is *not necessary* and *yet meaningful*. Unity of necessity and contingency, of essence and appearance, is the 'stuff' our world of experience is 'made on.'

notion of e.g. topological space has characteristics in common with the original Euclidean structure.

In an analogous way a living organism can be seen as an individual member of a species, as an element in an ecosystem, as a colony of cells, as the implementation of a morphogenetic field in chemo-physical processes, etc., all different structurings with different individual substances and different properties leading to different theories. Yet all attempt to express the essence of the same phenomenon.

The actual structuring of a field of experience depends on the specific structurability of this field. Mathematical structures on the other hand are constructed and understood from the *general*, abstract principle of structurability. In a sense structuring precedes mathematics,[32] which is based on the discovery of its general principle. On the other hand mathematics depends on the actual structuring of the experienced world for its specific contents, which have been generalised step by step from early structures generated by counting and measurement. Therefore even the most abstract mathematical theories can be seen as generalisations of the arts of counting and measuring. The inspiration for these generalisations partly comes from experience with applications, partly from experience with symbolic constructions within pure mathematics. In information technology these two lines seem very gradually to merge into one.

1.2 Mathematics and mathematical thinking

In Section 1.1 the activity of mathematical abstraction was described. If the term is taken in a broad sense, it applies to four partial activities, which are of course usually simultaneous in time, being distinguished only by their different cognitive functions: analysis; structuring; conceptualising of structure and substantialising of structure. Structuring and substantialising a structure involve mathematical abstraction in *the strict sense*, i.e. understanding the principle of structurability. For structuring this is necessary, because the

---

32   Structuring can in a certain sense be regarded as the 'mathematical' component of experience. So it certainly precedes mathematics as a separate discipline, logically as well as temporally. Yet it can at the same time be said to be the 'mathematisation' of the world of experience. This mathematisation probably originated together with conceptual thinking itself. Logically it involves the conceptualisation of properties and it is also presupposed by it. The division of the world of experience generates both structures and predicative concepts. As these are expressions of the human *intellect* they cannot be based on an *arbitrary* division. Clearly there is something between arbitrariness and strict necessity: *meaning*.

possibility of structuring must be understood in actually doing it, not of course theoretically beforehand, but practically *while* doing it. Substantialising a structure, i.e. conceiving of it as a quasi-substantial mathematical object, can only be performed if one understands the principle of structurability as opening a perspective on an ideal world of mathematical objectivity.

Whereas in mathematical thinking this last operation is not necessarily explicit at all, we nowadays[33] use the word 'mathematics' for the explicit activity of constructing and studying mathematical objects. Retrospectively we are able to distinguish this activity in all disciplines called 'mathematical' in the past, even in times in which mathematics and natural science were not clearly distinguished from each other. In the present section, mathematics proper as well as mathematical thinking in general will be discussed and characterised.

Mathematics is the activity of constructing and studying mathematical objects, i.e. ideal structures. Although ideal, these structures are *objective*, in the sense that they are not arbitrary combinations of properties but meaningful unities which can be investigated. That of course does not prevent them from being at the same time 'subjective' in the sense that they are given only for a knowing subject. Mathematical objects have both the character of objectivity and the character of subjectivity, in this sense, in common with physical objects. The difference is, indeed, that mathematical objects are *ideal*, which means that they do not exist physically. How they then *do* exist is a classical problem of the philosophy of mathematics, which will be discussed in Section 1.3.

What concerns us here is the question of how progress is made in mathematics. Progress means increase of knowledge and deepening of insight. But the question is: insight in and knowledge of *what*? We might answer: of mathematical objects; but this is circular, for we have abstained from the question how and even whether these objects exist. The only epistemological

---

33 In the 18th century the term '*mathematica*' denoted a whole range of disciplines, from traditional arithmetic, geometry, theory of music and astronomy to technical specialties such as the design of fortresses and ballistics. In the 19th century there was a very gradual development of the acceptance of the idea of 'pure mathematics'. Nowadays the general identification of 'mathematics' with what was formerly called 'pure' mathematics is illustrated clearly by the use of the term 'applied mathematics' for disciplines in which the application of the results of 'mathematics'(=pure mathematics) is studied. The term 'mathematical modelling' or 'mathematisation' seems nowadays to be used for forms of mathematical thinking involving implicit as well as explicit mathematical abstraction within the same context.

commitment we have made so far is that there is a principle of structurability, which is a *real* principle. If that is so, it is possible to gather knowledge about this principle and, what is more important, about its consequences. A rather popular book about mathematics in the 1950's was called: *The Consequences of 1=1*. I am inclined to agree with that description of mathematics, provided 1=1 is read as a shorthand for 'the principle of structurability.' The word 'structurability' indicates a *potency*. It means that things are able to possess structure, that they are divisible in thought into individual constituent parts, and that they may be considered as 'built up' from these parts, somewhat as a house is built up from bricks.

Common sense, as well as many scholastic philosophers, associates this property of structurability with being *extended*. For if we imagine something built up this way, we imagine it as extended in some space. But this only makes it plausible that extension is *sufficient* for structurability. But is it *necessary*? Time e.g. seems not to be extended, yet it is structurable. Henri Bergson concluded from this that what in modern science is named 'time' is really a spatial image of time. But this begs the question, for what enables us to make a spatial image of time? It must in that case have something in common with space, and that something might consist precisely in its structurability. If we look, on the other hand, at the representation of extension by the mathematical continuum, we see that, by considering this continuum as a set of points, we have abstracted from the fact that it is extended! For the same point-set may be given the topology of a 0-dimensional discontinuum,[34] which does not contain any extended part. So extension is only reconstructed by a topological specification of the notion of 1- (or n-) dimensionality. The mathematical object is not *really* extended, just as a point does not *really* move on the time-axis. Extension and motion are only *represented* by certain mathematical objects.

Now if structurability is a potency, the only way to investigate it is by its actualization, i.e. by structuring, as it is involved in what we have called 'mathematical abstraction in a broad sense.' That means that some element of our world of experience inspires us to think of an abstract structure which could be an actualization of its structurability. So, for example the possibility of moving ourselves and the things around us, gives rise to the structure of Euclidean space as one of the possible actualisations of the structurability of

---

34   This is what is termed the Cantor-discontinuum, which is obtained as a limit of the process of removing the middle third from every interval obtained by the previous step. So one starts e.g. with the interval [0,1] (I_1), the first step gives $I_2 = [0,1/3]$; $I_3 = [2/3,1]$; the next step gives 4 intervals; and so on. The limit-set can be characterised by the formula:

$$\sum_0^\infty 2 \frac{a_i}{3^i}$$

,where a_i is 0 or 1. This means that there is a 1 to 1 correspondence between the points and the set of subsets (powerset) of the set of natural numbers just as in the case of the interval [0,1] itself.

our world of experience. The question whether there are *preferred or preferable* actualisations will be dealt with in chapter 2. The possibility of counting things and comparing their numbers gives rise to the structure of the natural numbers. These were the only ways in which mathematical objectivity was generated in antiquity. In the modern era a completely new field of experience was added to the 'natural' experience of space and number. Laws of change and motion and equilibrium were discovered, giving rise to new and complicated mathematical structures, such as infinite sequences and series, real and complex numbers, rings and fields, and so on. When in the nineteenth century mathematics was emancipated from natural science, it was not only enriched by these structures in comparison to the mathematics of antiquity, but it had developed a completely new way of generating such structures by itself. In order to be able to explain this, a few things must first be said about mathematical notation.

Mathematical disciplines need ways of denoting their objects. This denotation is a kind of *naming*, but not in the same way as we name objects in our world of experience. This latter kind of object may be given to us by sensory perception, and the use of the names of such objects is not independent of this. Naming and perceiving are interwoven in human practice. But mathematical objects cannot be perceived. Of course, by definition they may be subsumed under certain concepts, but in principle there are infinitely many objects answering to the same conceptual definition: natural numbers, sets of three elements, points in space, etc. In their individuality these objects remain undetermined. Therefore we can only name them by *postulating* their individual identity, and this we do precisely by *naming* them. We say: let P be a point on line l. Now as long as we do not postulate the contrary, the individual identity of this point is represented by the 'Gestalt'-identity of the letter P. Even stronger: every time we use this letter as a name of this point, we silently postulate that it *is* the same point we mean. Mathematicians may easily follow a text in which this letter has two or more completely different meanings, understanding from the context which identification postulate has to be made each time.[35] So the first function of mathematical naming is to *postulate* the individual identity of mathematical objects.[36]

---

35   Perhaps it is for this reason that in the well known joke a mathematician is defined as a person who says A, writes down B, means C, while the right answer is D.

36   Saul Kripke introduces in his *Naming and Necessity* the notion of 'rigid designator' He does not seem to notice, that the human practice of naming something in the world of experience *excludes* by its nature the existence of rigid designators, for this practise involves the fallibility of the identification of individuals. The only practice in which we *do* have rigid designators is in the naming of individual *mathematical objects*, for there

From elementary geometry we are familiar with another function of mathematical notation, although we may not have regarded the drawing of *figures* as a form of notation. Yet it is, for the drawn figure is *not* the geometrical figure we mean by it. It is only a sign or symbol for the geometrical figure. Yet it does more than arbitrarily name an individual object. It *has* a certain structure, which in a certain sense is the same as the structure the geometrical figure *is*. 'In a certain sense,' for a real structure cannot coincide with a mathematical structure. They only have *structural properties* in common. A drawn triangle denotes a geometrical triangle, three fingers stuck out denote the number three, 'a_1, a_2, a_3,.....' denotes an infinite sequence. In fact, we think of what is used as a symbol as structured according to the mathematical structure it denotes. There is an immediate agreement between the structure we give to the sign and the structure designated. This is the most natural kind of mathematical symbol,[37] so natural, in fact, that it had to be *discovered* by the ancient Greeks that it was only a symbol, that a drawing did not *prove* anything. With that discovery came the practice of using arbitrary, *conventional* symbols, such as letters.[38]

But there is a third element in mathematical notation, for which the decimal notation may stand as an example. This notation, say for natural numbers, in on first sight conventional, for the numerals may be regarded nowadays as conventional symbols. But their *combination* is not completely conventional. It has a certain structure, which makes it convenient to use the symbols, but which is not the structure of the natural numbers themselves. Decimal notation is a *notation system*, consisting of a certain use of conventional symbols which involves their being combined into expressions which are convenient for denoting natural numbers.

The three elements of mathematical notation mentioned here have a systematic coherence.
a) Immediate representation of the 'geometrical figures' type makes use of the structure we see in the sign to indicate the type of structure denoted by it. But both structures - of the sign and

---

we can have no doubt about the identity of an individual: we either postulate it to be absolutely the same as an individual mentioned before, or we postulate it to be another.

37   This kind of 'immediate' symbolic representation is used by all working mathematicians as the use of visual or imagined 'pictures' of the objects they are dealing with, and their positions in enveloping structures. Those pictures still do not prove anything, but they form an indispensable heuristic tool.

38   The practice of using letters for *numbers* may be older than that, but probably these letters were not experienced as *arbitrary* symbols.

of the designated object - are not explicitly distinguished and are therefore linked together rigorously and unconsciously.

b) Conventional representation is opposed to this immediate linkage. It is based on the arbitrariness of the relationship between sign and designated object. The only essential linkage is that the individuality of the sign is used to symbolise the individuality of the designated object. This is so to speak the minimal remainder of the immediate representation in the conventional representation.

c) Systematic representation makes use of the fact that signs can have a structure of their own, but not in order to indicate the nature of the designated object, but to handle it more conveniently. The relation between the system of designated structures and the system of signs has itself become conventional, but also structural. There is a mathematical relationship between both structures as soon as the system of signs is mathematised. **(c)** unites **(a)** and **(b)**.

Now a system of notation may have a structure of its own, which, if mathematised, i.e. considered as an ideal structure, might be of a nature different from the original structure for which it was a notational system. A polynomial, for example, was originally a notation for a sequence of arithmetical operations. But by their structure, polynomials of a certain type can be idealised and considered as elements of a polynomial ring, which is a new mathematical structure, linked in a mathematical way to the original number system. Another important example is the system of complex numbers. The algebraic notation system gave rise to certain terms, such as the square root of a term denoting a negative number, which had no meaning in the structure of real numbers. But the coherence of the notation system gave rise to a new number structure in which these terms had a meaning. In this way many kinds of extensions of existing structures were generated, for example all kinds of extensions of number-systems and the compactifications in topology.[39] As a last example I mention category theory, which was developed from the notation using arrow-diagrams in order to indicate functional relationships among algebraic or topological structures.

We can conclude by distinguishing three ways of generating new structures in mathematics:

---

[39] Here, instead of using certain object-like notations like $\sqrt{-1}$, notations for *functions* or *sequences* are used in order to generate new objects. One can for instance embed the natural numbers in a space consisting of the totality of all infinite sequences of natural numbers. In this space every infinite sequence of elements has a limit point. It is *compact* in the topological sense. It is easier however only to include increasing sequences and identify them all. The effect of this is that a point 'infinity' is added to the set of natural numbers. Between the first 'maximal' and the last 'minimal' or 'one point' compactification, there are many other possibilities.

1) By inspiration from the direct structuring of aspects of the experienced world. This method this is still in full swing today in mathematical computer science.
2) By inspiration from the notational systems of mathematics itself. This has led in our century to the development of mathematical logic and metamathematics.
3) By generalisations of existing structures. This has produced branches of mathematics such as group-theory and other parts of abstract algebra, set theoretical and algebraic topology, set theory itself, many systems of mathematical logic, etc.

After the emancipation of mathematics from natural science in the nineteenth century, these processes led to the incredible riches of twentieth century mathematics. Yet it is good to realise that these riches have not been created *ex nihilo*, but by a further development of mathematics as it has grown from early times. Its content is never an *arbitrary* actualization of the principle of structurability. All mathematical structures we know today are linked together in a body of mathematical practice and knowledge. They have a meaningful place in exploring the realm of ideal structure, which produces knowledge, because structurability is based on a *real* principle. One can nowhere draw a line between *classical* mathematics consisting of arithmetic and geometry and '*modernism*' leading up to ever more abstract structures, nor is it meaningful to dismiss certain very abstract parts of mathematics as irrelevant for actual mathematical knowledge, or even as based on misunderstandings of the true nature of mathematics.[40]

---

[40] This is what some neo-scholastical philosophers of mathematics tend to do. They tend to distinguish the more traditional disciplines of arithmetic and geometry from the more 'abstract' branches of mathematics, and to deny the status of 'knowledge' of the latter. As if Plato has not already shown that even the number two is alarmingly abstract. Such a tendency also existed in the beginning phases of 'intuitionism', though the intuitionists drew the line between the 'constructive' and the 'non-constructive'. Nowadays, however, most 'intuitionists' have accepted that non-constructive methods can also have a mathematical meaning. Another branch of mathematics in which some mathematicians shrink from highly abstract entities is axiomatic set-theory. It is considered questionable whether the existence of inaccessible cardinals or even the validity of the continuum hypothesis has any impact on real mathematics. The same has been thought of Gödel's incompleteness theorems. All hesitation of this kind, in my opinion, has something to do with confusion of mathematical and physical objectivity. Objects such as the natural numbers are considered to be 'more real' somehow than for instance a high cardinal number. But in fact they are not, and Plato already understood that. We are only somewhat more accustomed to their practical use in structuring the world of experience.

All this concerns mathematics as an autonomous discipline, which in antiquity was and in the nineteenth century could again be distinguished from the sciences and technologies - developed since the 'scientific revolution' - in which mathematical thinking plays an important role. In such sciences and technologies, what I have called mathematical objectivity - implicitly or explicitly - plays the role of a tool in it which is used in order to structure the world of experience, the structurability of which must be already understood. Mathematical thinking here has the character of a *reflective*[41] motion of thought. Attention is directed from immediate experience to its possible structures, a process usually named *analysis* which is clearly described by Newton in the preface to his *Opticks*:

> As in Mathematicks, so in Natural Philosophy, the Investigation of difficult Things by the Method of Analysis, ought ever to precede the Method of Composition. This Analysis consists in making Experiments and Observations, and in drawing general Conclusions from them by Induction, and admitting of no Objections against the Conclusions, but such as are taken from Experiments, or other certain Truths. For Hypotheses are not to be regarded in experimental Philosophy. And although the arguing from Experiments and Observations by Induction be no Demonstration of general Conclusions; yet it is the best way of arguing which the Nature of Things admits of, and may be looked upon as so much the stronger, by how much the Induction is more general. And if no Exceptions occur from the Phaenomena, the Conclusion may be pronounced generally. But if any time afterwards any Exception shall occur from Experiments, it may then begin to be pronounced with such Exceptions as occur. By this way of Analysis we may proceed from Compounds to Ingredients, and from Motions to the Forces producing them; and in general from Effects to their Causes, and from particular Causes to more general ones, till the Argument ends in the most general. This is the Method of Analysis: And the Synthesis consists in assuming the Causes discover'd, and establish'd as Principles, and by them explaining the Phaenomena proceeding from them, and proving the Explanations.[42]

---

41   I use the spellings 'reflective,' 'reflection' etc. when a property or activity of the *subject* is intended; 'reflexive,' 'reflexion' etc. when those words characterize the nature of an *object*.

42   Isaac Newton, *Opticks*, Query 23/31 (1730 ed.) Dover reprint, pp.404-405

This description also refers to the opposite motion of thought, usually called synthesis, in which the resulting structure is used in order to *combine* empirical knowledge of its elements into an explanation of the phenomena.

This presupposes that the empirical knowledge is in such a form, that it is indeed combinable on the basis of a structure. When a motion or a force is analyzed by the well known parallelogram method, which is a structuring of a physical phenomenon using geometrical objectivity, the partial motions or forces must necessarily be expressed as continuous quantities. This means that they must be understood as *measurable* or, more in general structurable in a way similar to the structuring of the whole phenomenon which is implied in its analysis.

Therefore the method of analysis and synthesis asks for an ever more refined structuring of the experienced world. Once one embarks upon the vessel of mathematical thinking, using structures rather than empirical concepts in order to reflect upon experience, there is no way back. The distance from common sense can become so great, that the distinction between natural science and pure mathematics seems to disappear almost completely.

Yet we know that common sense always remains as a presupposition, in science as well as in technology. The methods and standards of measurement must be described ultimately with the help of ordinary empirical concepts and the same holds for the most elementary actions in technology. But the world of structures generated by this process, makes itself more and more independent of these basic standards and actions. Normalisation makes the latter common to all cultures and it seems that structure only becomes decisive. All measurements seem comparable and a program can be run on any computer. This may create the suggestion that we really live in the ideal world of mathematical objectivity. It is true that daily life seems to teach us otherwise, but cannot this be pure appearance? We speak of 'things', 'matter', 'qualities', but they can probably all be reduced to structures and why should the result of this reduction not be the *essence* of what is given as appearance to our common sense?

But if mathematical thinking is understood in an even more general sense as the use of structure in any practical or theoretical handling of the world of experience, it is to be found even in pre-modern practice. In this form it does not involve mathematical objectivity in the proper sense. Practices as counting or measuring or dealing with certain other schemes of action are structural in nature but do not necessarily involve the idea of *pure* structure. This conception comes in the neighbourhood of Heidegger's famous etymological investigation of the words μαϑησις and μαϑημα⁴³. According to Heidegger μαϑησις is what can be learnt and taught about a thing. That means that it can be known in absence of the immediate experience of the thing itself, it is the very basis of our recognition of this experience as the experience of this specific thing. But this independence of the immediate experience is, as we shall see in the next chapter, a characteristic of *all three* degrees of abstraction. Not only structure can be detached from the actual situation, but also concepts and principles. Heidegger starts his investigation by distinguishing what can be learnt about the *specific use* of a thing from what is more fundamental, and must be known already before this specific use can be mentioned at all. I intend to believe contrary to Heidegger, that mathematical thought has more to do with the - learnable and teachable - *specific* use of things, than with their more fundamental characteristics.

Mathematical thinking in the sense of the use of structure - in whatever degree of explicitness - is a form of *reflection*, in which experience is measured by means of the principle of structurability, and the result is amazing. Where does the seemingly unlimited power of this principle come from?

1.3 Philosophies of mathematics and of mathematical thinking

In the foregoing part of this chapter, the classical question of the mode of existence of mathematical objects has been left aside. Now three important types of answers to this question will be discussed. This is of consequence with respect to mathematism, for, as we already saw, in a Pythagorean metaphysics, mathematism should be considered as a very wise attitude, whereas in an Aristotelian metaphysics it is an intolerable reductionism. And the difference between different types of metaphysics, as far as mathematics is concerned, manifests itself it the corresponding answer to the above-mentioned classical question.

The three main types of answer[44] amount to the following:

A) The objects of mathematics exist in an ideal realm of being.

---

43   Martin Heidegger, *Die Frage nach dem Ding*, 1975², pp. 53-58.

44   They are inspired by Evert Beth's *De wijsbegeerte van de wiskunde van Parmenides tot Bolzano*, 1944

B) They do not really exist, but they are only thought of *as* existing in a certain way.
C) They exist as a stage in the development of reality from unity to plurality.

All three types may be specified in very different ways. **A** encompasses Fregean logicism as well as different types of 'Platonism.' **B** may or may not encompass Aristotle's view, but is certainly covers Hegel's conception of mathematical objectivity.
C covers neo-Platonism, e.g. Plotinos, but also Brouwer's constructivism.

The answer-types will now be discussed in such a way that it becomes clear that they are not independent of each other, but constitute different aspects of the same problem.

A) The objects of mathematics are not, like Plato's ideas, pure forms of which there is only one of the same kind. They are thought of as *individuals* within their kind. Therefore many of the fundamental problems we meet when we try to understand Plato's doctrine of ideas do not arise in the realm of mathematical objects. This may explain the tendency in the later Academy to 'mathematise' the ideas. And it probably explains why there are still many 'Platonists' in the philosophy of mathematics, whereas real full-fledged philosophical platonists are indeed of a very rare kind.

We really think of mathematical objects as if they exist in some other world, which is not accessible to our senses, but only to the understanding. Why should this way of thinking be deceptive? Many important thinkers have supported this point of view. It is doubtful whether Plato himself was one of them,[45] but the academians Xenocrates and Speusippos certainly were. In modern times the ideal world of mathematical objectivity was often identified with nature as it is in itself (distinct from nature as it appears to our sense-perception). In our time Frege, Russell, Cantor, Gödel and even a physicist such as Roger Penrose are 'Platonists.' The objections against 'Platonism' in the philosophy of mathematics come from two opposite directions.

First there are the objections in the line of Aristotle's 13th book of *Metaphysics*, amounting to the idea that mathematical objects are *too poor in content* to exist by themselves. They do not have qualities; they do not have an inner unity;[46] abstract elements seem to be more fundamental than

---

45   See note 18 above.

46   Leibniz uses this argument in his *Discours de la Métaphysique* (Discourse on metaphysics) against Descartes' *res extensa*: '...that the whole nature of body does not consist solely in extension, that is to say in size, figure and motion, but that there must necessarily be

complicated objects, whereas in an organism for example the complicated whole is more fundamental Moreover in mathematical objects parts and wholes seem to be equally substantial. On top of that, Aristotle argues, the fact that mathematical objects can be *studied* as if they were substantial does not imply that they are *in fact* substantial.

The second kind of objection is of a more positivistic nature. Such objections may be considered as applications of 'Occam's razor.' There is no need to assume anything above or outside the perceptible world in order to explain this world. So what counts in mathematics is its applicability, and Platonism is of no help in explaining this applicability. It concerns the objects of *pure* mathematics only.

Aristotle's argument is of a metaphysical nature, and we shall deal with it in Chapter 4. To the positivistic argument, a platonist would probably answer that Occam's razor is wrongly applied, because even if we can explain the *world* without assuming ideal entities, we certainly cannot explain our *understanding* of the world without them. If we can apply mathematics at all, its ideal objects cannot be purely arbitrary fantasies. It would be the extreme opposite of the positivistic attitude to say that the world is understood by means of poetry, and even poetry does not usually consist of *arbitrary* fantasies. So the perceptible world must have something to do with ideal entities. And Plato's solution, that they are *reflected* in this world, or that this world *participates* in them, is not superfluous, unless one has an alternative for it. The usual positivistic alternative is that with help of mathematics we *order* the data we gather about the world. It is argued that the application of mathematics does not add knowledge to a disorderly summing up of facts, but makes it more accessible, like the lexicographic order of a dictionary or a telephone directory. This argument, however, the Platonist would say, begs the question. *Why* does a mathematical order make the facts more accessible? The contrary is sometimes true. We all know the simple problems, solved easily by common sense, but difficult for mathematicians, who are looking for all kinds of structures of which an average person has no idea.[47] Moreover it is known in computer-science that data ordered at random are sometimes

---

recognized in it something which is related to souls and which is commonly called substantial form, ...', English edition p. 18.

47   The most stunning example is that of a bar of chocolate consisting of four rows with ten pieces in each row. The question is: what is the minimum number of times of breaking a piece in two (never two pieces together) required to separate all the pieces. Mathematicians even used a computer in order to find out the (obvious) answer.

more accessible than data ordered according to a pre-established principle.[48] In other words, the accessibility of facts is also a fact, and the problem of the intelligibility of the world is not brought nearer to its solution by stating that certain facts have a subjective side. For *all* facts have a subjective side, but also an objective side. The dictionary is *objectively* more accessible than a random word list for persons possessing knowledge of the alphabet and its use. And of course one has to *know* mathematics in order to apply it. But that was just the question: what does it mean to know mathematics.

Yet there is an element in the positivistic argument which is not properly countered by the above reasoning, and which is akin to Aristotle's criticism. The participation of the perceptible world in something of an ideal nature does not presuppose the *separate existence* of ideal objects, but only the *independent determinateness* of the ideal. The word 'independent' is meant here with respect to the perceptible, factual world. A society may for instance participate in the ideal of justice, but this does not presuppose that somewhere, in heaven or on earth, a perfectly just society exists or even that it *could* exist.[49] It only presupposes that the ideal of justice *has a definite content*, independently of the real state of affairs, so that this state of affairs can be measured and judged by it. And of course we should be able to *know* this content *as the content of an ideal*, independently of the factual situation. With regard to mathematical objectivity this objection to 'Platonism' marks the transition to position **B**: the theory of abstraction.

B) I do not claim here to give a correct interpretation of any specific philosophy from the past. As I did with 'Platonism' I shall try to present the position in a form which seems to me to be the strongest and most defensible now. The same strategy will be followed for the objections against the position, so that the 'inner' motion of thoughts comes out most clearly.

Mathematical objects, then, are thought of as ideal, quasi-substantial structures. This means that when their substantiality is pressed too much, the whole idea evaporates into nothingness. We live on quicksand, but that is no

---

48 For instance the well known problem of the travelling salesman is solvable in polynomial time by a non deterministic algorithm, but not by a deterministic one.

49 According to an old adage, *summa ius summa iniuria*: the highest justice is the highest injustice. For to suppose that justice is perfectly realised excludes any claim for justice inside this supposedly perfect realisation.

problem as long as we do not consider ourselves too weighty.[50] Mathematical thinking is based on a certain amount of intellectual abstinence, which is often felt by mathematicians. Mathematical thinking is walking on an upland plain, not climbing a mountain peak. One has to exert oneself to reach the level of mathematical abstraction, but once that is mastered, one is equipped to attack any mathematical problem. Of course one has to find one's way in the specific field, but that does not introduce new *fundamental* difficulties. This may have to do with the fact that there is only one fundamental intuition involved: the grasping of the principle of structurability or of *ideal, intelligible matter*. On the basis of this principle, inspired by whatever material we may encounter in the practices of technical invention, symbolic representation or generalisation,[51] we can construct and investigate as many fields of mathematical objects as we want. We can always find axiomatisations for them and prove interesting and often applicable theorems. Mathematics is at the same time an art and a science, for we create structures in thought and investigate these creations, discovering that they are not arbitrary at all, but all specifications of the same real principle.

Furthermore, this is precisely the reason for the applicability of such structures, for it is according to this same principle that we structure our world of experience, in theory as well as in practice. It is a metaphysical question where this structurability comes from, and an epistemological question what power enables us to understand it. But that does not make its status less clear: structurability is understood by an act of abstraction, in which we concentrate our attention on the structural aspect of all reality and neglect in thought all 'sensation' and 'emotion' (these words are to be taken literally to mean: experienced quality and experienced change; but their colloquial meanings also have something to say about mathematical thought!). The understanding of this principle is what makes technical mastery of the world possible. Because by re-structuring, by recombining the elements and laws of nature, we can

---

50 This pun is borrowed from H.M.J. Oldewelt, Professor of Philosophical Anthropology at the University of Amsterdam from 1946 till 1967. Of course he did not refer to mathematical objectivity, but to the human condition in general, for which, in my opinion, it is less true in a philosophical sense and yet wise in a practical sense. It may also be applied to Sir Karl Popper's idea of 'piece-meal engineering' in science, which I take to reveal a mathematistic tendency in his thinking.

51 These different sources of inspiration are not equally valued by all adherents of 'abstractionism'. Therefore it is sometimes difficult to recognise them as belonging to the same mainstream. Hilbert, for example, stresses symbolic representation strongly, whereas Dingler and Husserl have an eye for the mathematical element in technical practice. For Aristotle, on the other hand, the theoretical universe of geometry was the first phenomenon to be explained.

obtain certain effects, without ever having to oppose a natural force, which of course would be beyond our powers, for we cannot change the laws of nature. Structure is always in a certain measure *indifferent* to nature, and it is this indifference we use to gain power, not over 'nature' herself, but over certain effects of processes which were taken to be 'natural' beforehand.

The explanation of the applicability of mathematics has always been the strong side of the theory of abstraction. It is really a 'down to earth' philosophy of mathematics, so we need not be surprised that its difficulties are of a metaphysical nature.

Where does such a principle of structurability come from; how is it connected to and distinguished from the material world? How is it possible to conceive of an 'ideal matter' by abstraction? In Aristotelian terms we might think of this principle as of 'the form of matter,'[52] but is this not a contradiction in terms? Is it not far easier and clearer to say that structure emanates from Unity itself, because of its immanent tendency towards self-distinction? With this thought, we leave the abstractionist domain and enter the neo-Platonic perspective of emanation.

C) *Consciousness* in its deepest home seems to oscillate slowly, will-lessly, and reversibly between stillness and sensation. And it seems that only the status of sensation allows the initial phenomenon of the said transition. This initial phenomenon is a *move of time*. By a move of time a present sensation gives way to another present sensation in such a way that consciousness retains the former one as a past sensation, and moreover, through this distinction between present and past, recedes from both and from stillness, and becomes mind.

As mind it takes the function of a subject experiencing the present as well as the past sensation as object. And by re-iteration of this two-ity-phenomenon, the object can extend to a world of sensations of motley plurality.

In measure of the irreversibility with which the subject has receded from an element of the object, this element loses its egoicity, i.e. gets estranged from the subject, and in measure of this estrangement, mind becomes disposed to desire and

---

52 Cf. Brigitte Falkenburg, *Die Form der Materie*, 1987.

apprehension, and consequently to positive or negative conative activity with respect to the element in question.[53]

This text brings us to the middle of the third type of philosophy of mathematics. It is more idealistic than abstractionism, and more dynamic than traditional 'Platonism.' Usually it is associated with the historical mainstream of neo-Platonism, but in modern thought it tends more towards subjective idealism. In this latter form, it regards mathematical objects as 'mental constructions,' which are independent of the world of experience. It is not denied that they help to structure this world, but the idea of a real principle of the world of experience, which is understood by abstraction, is rejected. We may intuitively have understood a principle of structurability, but this intuition must be purely spiritual, not only in nature, but also in *content*. If there were no material world, and we were pure, immaterial spirits, we could do mathematics as well or even better than we do it now. In the more objective, antique form of this view, even God, by letting plurality emanate from Himself, from the absolute unity, does mathematics. Perhaps the modern idea of God being a mathematician is a sign of neo-Platonic influences. All forms of this view regard mathematics as being a more *direct self-expression* of a fundamental spiritual unity than the material world is. Mathematics is closer to the absolute, closer to God, than material things or even living beings. Therefore neo-Platonism is the favourite philosophy of mathematics of those tending towards *mathematism*. It seems to imply that we do not need anything other than mathematics in order to understand and explain the world of experience. This way of seeing finds support in the indirect evidence[54] that Plato himself regarded the material world as a kind of degenerated mathematical objectivity. Matter tries to live up to the mathematical ideal, but it never succeeds completely. For the abstractionist this could be a sign that there must be *other* principles governing the behaviour of material things, but in 'emanative' thought it is only a negative phenomenon, a στερησις or *privatio*, the absence of a perfection.

The most important objection against neo-Platonism is that *it gives no answer to the question of the mode of existence of mathematical objects*. Is it only the absolute unity that exists, or do all its emanations exist as well; and how is this existence distinguished from the Unity itself? Is it a deficient mode of being, participating in the unity, or an autonomous, created being? Or, in the more subjectivistic variant: does the spiritual subject *project* its world, or

---

53   L.E.J. Brouwer, 'Consciousness, Philosophy, and Mathematics, 1949, pp. 1235-1249

54   See note 18

create it; or does it exist independently? This ontological indeterminateness reflects itself in an epistemological dilemma: what is the status of mathematical knowledge? Is it indirect knowledge of the absolute unity, and therefore metaphysical knowledge in disguise? Or is it only knowledge of a deficiency, and therefore not knowledge at all? These are the only consistent positions. All views between them tend to become either Platonistic or abstractionistic. For the deployment of the unity, if it were admitted to exist in a positive way, would either be an ideal world or an independently existing - but possibly created - real world. Or, to return to Brouwer's text: in his way of seeing, knowledge of the natural numbers is either self-knowledge of the mind, or knowledge of a fata morgana, and therefore no knowledge at all. If we do not accept one of these possibilities, the natural numbers are either ideal entities or abstracted from a real world.

1.4 An attempt towards synthesis

The three types of 'philosophy of mathematics' mentioned above each have an unsatisfactory element. Nevertheless, all three are also perfectly correct in some respect. Platonism is right about the ideal nature of mathematical objects, abstractionism is right about their relationship to the world of experience, and emanatism is right concerning the dynamic character of their generation or construction. This may give us the feeling that these perspectives do not exclude each other essentially. In this section I shall try therefore to unite these perspectives into a coherent philosophy of mathematical abstraction and objectivity. Such an attempt will unavoidably look very 'old-fashioned,' as it is bound to use classical terminology in order to indicate in which respect it is a synthesis of the three classical positions. But then we see these positions reappearing in all non-superficial attempts towards a philosophy of mathematical thought. So it may be best to begin at the beginning in trying to unite them. Moreover, old-fashioned terminology might still be useful in expressing an original thought.

From abstractionism we borrow the idea that the *principle* of all mathematical structure comes from the world of experience. This does not imply that this principle is 'empirical' or 'contingent.' The property of structurability belongs to the essence of the mode of being of this world. This mode of being is commonly called 'material,' which means that its determinations are never complete. Every material thing is capable of change, and is therefore not completely determined; it contains an element of indeterminateness, usually called 'matter.' This indeterminateness is the cause of changeability and structurability. Aristotle considers matter to be 'passive potency,' which means that it does not change by itself, but is changed only by some active cause. This explains at the same time both the permanence and the change of material things. I do not see how one can conceive of a real material world without supposing such a principle of passive potency. Of course, when we actually distinguish 'form' (=determination) and 'matter' (=determinability) in a material reality, this distinction is only *relative*. What we have called 'matter' is known to be *a particular kind* of matter and therefore again understandable as consisting of 'form' and 'matter.' This is precisely the reason why Aristotle calls form and matter *principles*. Only in their *ultimate* content are they independent of any particular case in which they are relatively opposed.[55] 'Form' in a concrete case is the form of a particular matter, and 'matter' is the matter of that particular form. But in order to make sense when we state this relative opposition, the opposites must mean something *independently of their opposition in a particular case*, otherwise the opposition could not be e.g. one of *matter* and *form*; they could not have a general meaning. It is difficult to think of these general meanings - determination and determinability - without imagining a concrete case, in which they are relatively opposed. Yet we can understand that such an opposition presupposes the specific nature of the general meanings. So I do not say that the principles *exist* independently; they only have an independent *content*. But this means that the content is in a more or less Platonic sense *ideal* with respect to the real, concrete unity.

Now, in so far as this independence of content exists, we, following Aristotle, speak of *principles*. They are united in a relative opposition in any reality governed by them, but they are not completely defined by this concrete opposition. They have a contribution of their own to the nature of the concrete thing and to the nature of the opposition in this concrete thing. This

---

55 Aristotle distinguishes in his book on Categories four types of opposition: contradictory; privative; contrary and relative. The opposition between affirmation and negation *of the same content* is called contradictory, e.g. being and not-being. Privative opposition exists between the possession of an essential property and the lack of it, like seeing and blind. Contrary is the opposition between extremes in a category, e.g. warm and cold, everyone and no-one. The opposition between correlatives as such is called relative, e.g. parent and child, whole and part. A relative opposition is present in an *absolute* sense, if the correlatives are completely determined by the relationship, like parent*hood* and child*hood*, subjectivity and objectivity. This is also the nature of the opposition of determinacy and determinability in a concrete entity.

contribution presupposes that they have an independent content, in which they may also be opposed to each other, but in a different way.[56]

Structurability is *ideal determinability*. How can we distinguish real and ideal determinability? Can we also distinguish real and ideal determinateness? Yes, if we look at the *concrete* thing, we can *imagine* that its determination and determinability are quite different elements, although in reality they cannot be separated; that they are completely indifferent to each other. We *can* imagine this, because of the ultimate, ideal meaning of the principles, but at the same time this is an idealisation, because in the real concrete things the principles are not independent, but united and relatively opposed within this union. In the perspective of this idealisation, determinability becomes structurability and determination becomes structure. A reality is structurable in so far as it is possible to understand its principles of determination (form) and determinability (matter) *in their ideal sense*, so that the *particular* form and the particular matter are quite indifferent with respect to one another. This means that the form is thought of as a form which can exist in *any* matter. It is only related to the *idea* of matter, which can be thought of as an ideal, intelligible matter. And, on the other hand, the matter is thought of as capable of having *any* form, as being deprived of all qualities of its own. Determination and the existence of this determination have become completely indifferent to one another. But this means that the determination can only exist as a way of arranging individual elements, which are also indifferent towards one another. Only in this case the determination does not concern the very nature of the matter it determines; anything could be arranged so as to satisfy the determination. In this case there is nothing in the elements preventing another arrangement and there is nothing in the arrangement preventing its being applied to other elements. If matter and form are *indifferent* to one another, the elements of their union must be mutually *external* too. In other words, the concrete whole must be understood as *structure*.

Of course structure in this sense does not really exist. Nevertheless, it is an *ideal possibility*, which is understood as such by abstraction from the *concrete* union of matter and form. Therefore, structure can also always be understood as *constructed* in an ideal way. It exists in a 'fantasy' going beyond material reality, based on insight into the ideal character of the principles of this reality.

---

56   This distinction between the ideal content of principles and their real or concrete presence is of central importance in the following chapters, and it will receive a more elaborate treatment there.

On this conception the Platonic element consists in the ideal character of the principles; the abstractionist element is the abstraction from the concrete unity of these principles in a particular reality; and the neo-Platonic element is the ideal construction of structures on the basis of insight in the ideal principles. Mathematical objects are indeed mental constructions, but these are based on the abstractive intuition of ideal principles from the concrete material world.

1.5 Mathematical thinking and logic

It is generally assumed that mathematical thinking is the most 'exact,' the most 'logical' form of human thought. What is this mathematical exactness, and what has it to do with logic? In line with these questions is another philosophical issue of importance: the nature of mathematical logic and its relationship to classical logic. All these questions will be touched upon in this section.

Why is mathematical thinking 'exact,' and what does exactness mean here? At first sight the answer is obvious. Mathematics makes all presuppositions and concepts it uses explicit, and reasons from these presuppositions exclusively according to the rules of logic. So every mathematical theorem can be considered as a complicated tautology, saying: 'This or that state of affairs is a logical consequence of the axioms of the theory.' A problem with this conception is that it identifies mathematics with deductive reasoning in general, so that in principle there is no difference between mathematics and logic. Yet we know from experience that deduction and exactness are not the same thing. In daily life and in empirical science we usually have many implicit premises in our deductions, making them rather misleading sometimes. But even if we could succeed in making these premises explicit, there would still be a problem. For what has this reasoning from general premises to general conclusions to do with the real particular elements of our field of interest? Might not this real field of interest change its rules while we are busy reasoning about it? Are we allowed to say that our conclusions apply to the same reality, the same individuals as our premises? How do we know that reality does not change the rules of the game while we are reasoning, like the Red Queen in *Alice in Wonderland*? How do we identify the individuals of our field? Can they *appear* to be the same on different occasions, but in reality be different, or the other way round?

There are two possible attitudes towards these sceptical questions:

The *first* is to admit that in empirical questions, reasoning can never be completely exact. We cannot be rid of all our silent presuppositions, and we have to test our conclusions in practice in order to find firm ground for them.

The *second* attitude is to say that we are not really reasoning about our primary field of interest, but about a *model* of it, in which we suppose the ultimate rules to be fixed and the ultimate individuals to be identified and unchanging. Change can only be represented in such a model by a series (or a continuum) of different *combinations* of elements. In other words, the model has to be *mathematical*. Then we are able to say that with respect to the model, the reasoning is exact, but that there are always uncertainties concerning the applicability of the model to our field of interest.

This teaches us that, indeed, the perfect applicability of logical reasoning characterises mathematical thought, but also that there is a reason *why* this applicability is so perfect, namely the ideal character of mathematical objectivity. The rules of a domain of mathematical objectivity are not determined by empirical investigation, but by postulation. The same is true for the identification of individuals. In mathematical reasoning each individual is given a name, and if it is postulated that this name indicates the same individual on a second occasion, there is no room for doubt whether it is really the same or whether it has been changed or exchanged.[57] This is not so with empirical individuals, because their individual identity is only concluded from its effects. If there seems to be a continuous trajectory, e.g. in a Wilson chamber, it may be one and the same particle moving along this trajectory. But we are never sure. If a person has approximately the same fingerprints as someone who has committed a crime, we may guess that he is indeed the culprit, but his defender may argue otherwise.[58] Individual identity does not really *appear*, it is phenomenologically transcendent; and it is a metaphysical problem as to what reasons we have for nevertheless assuming its existence.

In more detail, this point plays an important role in the discussion about what is called the 'proof by exposition,' also known in the history of logic also as 'Locke's triangle.' The example Locke introduced was the proof that the sum of the angles of a Euclidean triangle is equal to two right angles. The proof is given by constructing through a vertice of an *arbitrary* triangle a line parallel to the opposite side. Now the problem is, what is to be understood by an arbitrary

---

57 Sometimes in mathematics we exchange individuals on purpose, in order to construct certain models - e.g. the models proving the independence of the axiom of choice -, but such models are based of course on a more fundamental model, assumed to be unchanging.

58 An interesting case in this respect was the process against *Demjanjuk*, who was identified by several victims as 'Ivan the Terrible,' a notorious concentration camp brute. The problem was that there seemed to be no way of really *proving* that he was the same person, however strong the suspicion might have been.

triangle? Is it acute-angled, obtuse-angled or perhaps even right-angled? It cannot be none of the three. It is an *arbitrary individual* which is thought of by abstracting from such properties. David Hume tried to solve the problem by suggesting that the arbitrary triangle is that triangle which is used as a counter example by someone who does not believe the theorem. But this begs the question, for we do not know beforehand of such a counter example, so this suggestion does not make the triangle we reason about less arbitrary. The only way to solve this problem is to point out that in mathematics individuality is *postulated*, so we can give individuality to an object *without specifying all of its properties*, and in fact we *always* do so in mathematics. For a mathematical object, say a group or a number, may be represented a second time by any of a variety of more highly specified objects. The number 3 for example, may be represented by a particular set of three elements, having properties which are not necessary for other specifications. In other words, the property of mathematical signs of being either a variable or a constant *depends on the context*. In mathematical logic this is reflected by the fact that every formalism has (infinitely) many possible interpretations and the semantic distinction between constants and variables is highly artificial.

So the fundament of the exactness of mathematics is the ideal character of its objects. This makes it possible to reason about perfectly identified individuals from perfectly established rules or axioms.

Now this situation is in a sense an *ideal* of logic, but it is not the *subject matter* of logic in general. Logic is the science of correct order of thought, not only in the ideal case, but in *all* cases. In the whole history of logic, however, the ideal case of mathematical reasoning has played a very important role. This has made logic an ambiguous discipline through the ages. On the one hand it has the face of 'formality;' from Aristotle's syllogism schemes to the twentieth century systems of mathematical logic, the discipline of logic seems to deal with exact rules, which invite one to use mathematical methods.[59] On the other hand there is logic as a 'practical discipline' concerned with clear thinking and correct reasoning. This is a far more casuistical discipline, for which the formal rules are only of paradigmatic importance. In this form it nowadays goes by the name of 'philosophical logic.' The gap between the two seems only to be bridged by a fundamental conviction that it is not for nothing that they share the name derived from λογος, a word so full of meaning that it easily envelops both sides. This made it possible for profound philosophers like Hegel, Frege and Husserl to try to derive from their conception of λογος the proper content of the discipline of logic in such a way as to unite its two seemingly disparate branches. This touches upon the question of the relation of logic and metaphysics, which I shall deal with in Chapter 4. Here I shall only discuss the relationship between mathematics and mathematical logic.

---

[59] And this has also been done throughout the ages; by Lullus, Leibniz, Boole, Schroeder, Frege, Lesniewski, and many others.

If logic is understood as being the self-reflection of thinking under the perspective of its correctness, mathematical logic is the self-reflection of mathematical thinking under this perspective. According to this conception it should be a mathematical reconstruction of the ideal of mathematical thinking. We have seen that this ideal involves two elements:
1. Making explicit all conceptual and structural presuppositions; and
2. Reasoning about ideal structures consisting of elements, the individual identity of which is purely postulated by the use of symbolic representation.

Now in fact mathematical logic has precisely these characteristics. It represents mathematical theories as axiomatic formal systems (this is the side usually called 'syntax') and it interprets these systems in terms of ideal mathematical structures (this side is called 'semantics'). Moreover, this interpretation is itself understood as a mathematical structure, usually a function or a class of functions. The postulation of individuals is also reflected thereby, for the elements of the formal systems are interpreted as names (definite or indefinite, i.e. constants or variables) of objects or relationships in the mathematical structure constituting the domain of the interpretation or - in the case of 'logical symbols' - in a more comprehensive structure in which the original interpretation-domain may be embedded. In the most elementary case of the first-order predicate calculus, the domains of interpretation are relational structures, and the interpretation of the 'logical symbols' is either given directly by a truth-definition or by means of a two element Boolean algebra.[60] I place the expression 'logical symbols' between quotes, because in a strict sense they are neither logical nor symbols. The elements of a syntactic structure are as ideal as the elements of any mathematical structure. The are not *used* as symbols, but only *thought of* as symbols, just as the natural numbers are not actually used for counting (the *numerals* are!), but only thought of as standards of cardinality or ordinality. A symbol however is by definition something *used* for designating something else. The 'logical symbols' are strictly speaking not logical either. They represent *mathematical* functions or functionals - Frege spoke of *truth functions*[61] - the values of which are again syntactical

---

60 For calculi of higher order the relational structures may be multi-typed; for quasi-operational systems such as the lambda-calculus a topology may be involved; for non-standard interpretations the two element Boolean algebra may be replaced by larger Boolean algebras or even lattices. In all these cases the fundamental scheme of syntax and semantics is maintained.

61 For Frege the value of 'F→G' e.g. is a function of the values of 'F' and 'G', and these values are either *TRUE* or *FALSE*. In modern mathematical logic the values of 'F' and 'G' are syntactical 'expressions,' and → is a two-placed function between such

expressions. It is of course true that their *interpretation* may be intended as a representation of certain logical connections. But this is a *mathematical* representation, and the fact that it may be varied - e.g. in non-standard interpretations - makes it clear that the representation cannot be expected to cover the original perfectly.[62] And, of course, it *presupposes* the original connection, for this is used in our epitheoretical[63] reasoning about formal systems, and even in our descriptions of the interpretations of the 'logical symbols.'[64]

Mathematical logic is an adequate self-reflection of mathematical thinking, for, as we have seen, its *content* as well as its *method* is mathematical. The most important result therefore is the kind of result we should expect from an adequate self-reflection of a form of thinking. It shows us the *limit* of this kind of thinking. The incompleteness theorems of Gödel and Tarski imply that all formalisations, syntactical or semantical, presuppose a point of view *not itself included in the formalisation*.[65] Of course this

---

'expressions'. The same remark of course holds for the word 'expression' as was made in the main text about the word 'symbol'.

[62] A clear example is *negation*. In a combinatory complete system there can be no proper negation, because the fixed point theorem states that every operation must have a fixed point, whereas negation by definition cannot have a fixed point. Russell's antinomy may be interpreted as an example of what happens if certain fixed points are constructible in a system. On the other hand it follows from Gödel's theorem that restrictions on the combinatory operations allowed in a system are always mathematically arbitrary. So in a restricted system, such as the first-order predicate calculus with limited term-forming, there are always meaningful expressions not allowed in the system, for which negation would make sense. So the scope of an operator representing negation is necessarily either too large or too small to cover the *concept* of negation adequately.

[63] This term is coined by H.B. Curry for reasoning in a language we in fact *use* to mean syntactic entities and properties. The usual terminology using the prefix 'meta' (metamathematics, metalanguage, metatheory, etc.) is ambiguous, for it does not indicate whether we mean the *actual use* of terms, or again only syntactical relationships of a structural nature. Curry distinguishes *U-language*, which is actually used, and *O-language*, which is the object of some mathematical theory. Cf. H.B.Curry, *The Foundations of Mathematical Logic*, 1963, pp. 28ff.

[64] If we describe e.g. the interpretation of $\rightarrow$ by a truth-table, a line of the truth table may mean that F$\rightarrow$G is true *if* both F and G are true. This description uses 'if' but it is not circular, for it is an element of the definition of an interpretation which establishes $\rightarrow$ as a mathematical *representation* of the use of the word 'if' in mathematical reasoning.

[65] This even holds for those 'perverse' formalisms which by their intricate 'infinitistic' structure escape Gödel's or Tarski's original theorems.

insight is also present in Brouwer's, Wittgenstein's and Curry's criticism of metamathematics in its pretension to produce a *foundation* of mathematics, but in the works of those thinkers it is only present as an intuitive insight, not produced by self-reflection. The self-reflection reveals something more than the insight in itself. Just as Descartes' self reflection resulting in his *cogito sum* reveals that the self-reflecting subject by this act of self-reflection *distinguishes* itself from its external world, so the incompleteness theorems reveal that mathematical thinking by its logical self-reflection distinguishes itself from its ideal world of mathematical objectivity. By understanding itself as the syntactical manipulation of external symbols and the semantical relating of ideal structures, mathematical thinking reveals that, in reality, it makes itself possible by *systematically looking away* from what it is doing. It conceives of ideal structures representing meaningful aspects of the world of experience, but it looks at these structures as self-sufficient quasi-substances in order to understand their ideal properties. It *projects* the whole world onto the perfect screen of mathematical objectivity, systematically looking away from the source of the projecting light: the intuition of the principle of structurability.

1.6 Mathematical thinking and technology

How do we now explain that mathematical thinking is so fruitful for technology? Why does technology gain by exact measurement and calculation, whereas at the same time too much precision and perfection leads away from its goals because of decreasing efficiency? Both mathematics and technology are characterised by inventiveness and ingenious constructions. In technology invention is intended to be in the service of some human practice, whereas in mathematics it serves knowledge of a structural universe. Yet in both cases it serves to find structures or constructions having certain preconceived properties. In mathematics these constructions serve as 'expositions,' generic individuals linking together two general propositions such as the parallel postulate of Euclidean geometry and the theorem concerning the sum of the angles in a triangle. In technology the constructions link together laws of nature and technological aims, like the hammer links together the law of the conservation of momentum and the aim of driving a nail into a wall. These examples also reveal a difference. In technology the laws need not be explicitly and quantitatively known, whereas in mathematics we always deal with explicit theorems. Clearly mathematics is most akin to a specific form of technology, i.e. 'scientific technology, in which we depart from explicit knowledge of natural laws and calculate effects before we try them out. And of course this is a very *mathematical* form of technology. We may say that in handcraft-like technology this mathematical element is *still implicit*, and that in technology based on information processing it is becoming *implicit again*, for we need not know exactly what is going on in the computer, that is precisely what we have it for. Certainly, in this latter form of technology on some higher level of design we still use explicit calculations and mathematical methods, but only in order to make them superfluous on lower levels.

The three forms of technology mentioned above are not arbitrary. They are developed in a theory about technology which is very useful for clarifying the relationship of mathematical thinking and technology.[66] I shall give a very short account of this theory here, accentuating those aspects which are informative for our subject.

Technology, like mathematical thinking, is based on a *principle*, called by Hollak the *technical idea* and formulated thus:

The technical idea is that abstract form of understanding by which mankind expresses its power over nature, practised by combining its forces in an original way into a new procedure.

Now we may immediately guess that this 'abstract form of the understanding' will have something to do with mathematical thinking. But we have to be careful on this level of philosophical investigation and not to identify meanings which only *seem* to be the same. In its context, 'abstract form of understanding' means the characteristic feature of *human* technology i.e. that an invention is understood as the realisation of a *concept*. That implies that such an invention is not restricted to the situation in which it was needed and applied, but that it is *univocal, universal* and *objective*. It can in principle be applied anywhere and by anyone. It is this univocal and universal objectivity which is indicated by the word 'abstract.' We see a similarity and a difference when we compare this with the way in which mathematical objects can be called abstract. They too may be applied anywhere and by anyone in the same way, but they have a feature lacking in the technical concept as such: their quasi-substantiality. They are thought of as *individuals* having an existence of their own, besides their applications. This difference accounts for the fact that mathematical thinking plays a different role in different forms of

---

66   Cf. J.H.A. Hollak, Betrachtungen über das Wesen der heutigen Technik, pp. 50-73; T.M.T. Coolen, Philosophical Anthropology and the Problem of Responsibility in Technology, 1987, pp. 41-65

technology. I shall now describe these forms and their connections more closely in order to analyze this role of mathematical thinking further.

Handcraft-type technology is the substratum of all technology, like vegetative life is the substratum of all life. It is always present, but it may be integrated in higher forms. On the other hand, the higher forms may already be anticipated in it. The computer is still an *instrument*, and the lock and key - already known to the ancient Egyptians - is already a kind of information processing system. A type of technology is thus characterised by two dimensions: the fundamental perspective of technical thinking and the typical artefact produced. Handcraft technology is based on the *accumulation of experience*, in which knowledge of natural laws is *implicit* and the concept is *immediately* connected with its realisation. This is the perspective of thinking which, as *know how*, is still present as a substratum in all technology. The typical artefact of this type of technology is the instrument or the *tool*. It is the first objectification of the technical idea: our interaction with nature is embodied in a contraption, which we place between ourselves and the forces of nature in order to handle them. The first instrument we use is of course our own body, and therefore many philosophers[67] have understood the essence of technology on the basis of the paradigm of the instrument as an extension of the body. It is a mistake, however, to understand *all* technology as instrumental. The instrument has two sides or 'interfaces': the side we handle and the side which interacts with nature. But by handling instruments, we still get blisters.[68] The solution for this problem is to let the instrument handle itself; or, more practically, to design a chain of instruments, one handling the other, possibly containing a loop which keeps everything in motion.

This is the idea of the *machine*. The design of such a thing requires ingenious deliberation. It usually does not come about by the accumulation of practical experience, but it requires *explicit* knowledge of natural laws, explicit construction of a concept and an accumulation of technical means to realise the concept. The machine is the artefact produced by *scientific* technology, leading to an *industrial* technological infrastructure, reflected economically in the accumulation of capital. This type of technology is characteristic of the modern era, which does not imply that there were no

---

67   This tendency is to be found in the work of Max Scheler, José Ortega Y Gasset, Arnold Gehlen and even Martin Heidegger, who in his characterisation of the world of 'durchschnittliches Dasein' as 'Zeugwelt' still accepts the instrumental point of view with regard to technology. Cf. J.H.A. Hollak, *Van Causa sui tot Automatie*, 1966.

68   "Ich erspare dabei nur der Quantität nach, bekomme aber doch Schwielen.", G.W.F. Hegel, *Jenaer Realphilosophie* Subjektiver Geist, B)Wille. (ed. Felix Meiner, 1969 p. 198)

machines before. The perspective of thought of this type is most akin to mathematical thinking. The concept is *abstract*, not only in the sense that it is universally applicable, but also in the sense that it has an ideal existence of its own. The inventor develops the concept in his mind and then tries to raise money to realise it. Of course in the process of realising the handcraft-like substratum comes up again, and some accumulation of experience is necessary in order to perform the realisation. But the main perspective is the one characterised by Heinrich Herz[69]: We make inner images of external states in such a way that the necessary consequences in thought of these images correspond to the necessary consequences in nature of the states themselves. In science such an image is a theory, and the correspondence is established by experiment. In technology the image is a concept and the correspondence is established by the realisation of the concept. Science and technology work in opposite directions, but they are both characterised by an explicit correspondence between an ideal and a real world. This correspondence itself may be identified with mathematical thought in general, and the exploration of the ideal world in itself with mathematics as a discipline.

Now, if we do not only *use* the correspondence principle, but also *understand* it, we may realise that it enables us to determine certain 'necessary consequences in thought' by relying on the correspondence, and reading them from the 'necessary consequences in nature.' In other words, we may use natural processes to simulate thinking. The perspective of *information technology* is thus seen to be a *self-reflection* of the perspective of scientific technology. A 'logical circuit' is not more logical in its design than any other electronic circuit, but it is called 'logical' because it is *used* for representing a logical feature of this design, e.g. a propositional connective. In general, a computer or any other automatic information processor is designed to be *used* for representing certain thought-processes. Artificial intelligence may, technically speaking, be a special subject within computer-science, but in principle *all* information technology is artificial intelligence. Its fundamental perspective is the self-reflection of the thought perspective of scientific technology and therefore of mathematical thinking. It does not correspond to mathematics, but to mathematical logic, from which discipline it has also borrowed its principal methods. Of course - again - this presupposes scientific technology. A computer is also a machine. And this presupposes handcraft-like technology, directly as well as indirectly. It takes experience to work with computers, it takes experience to make them, and it takes even more experience to produce software. In none of these cases is mathematical thinking or its self-reflection sufficient.

---

69   Heinrich Herz, *Prinzipien der Mechanik*, Vorrede

Now the question may be raised whether this self-reflective form of technology is likely to transcend itself into another, even more sophisticated form. Hollak holds the view that it is not. He argues that in this form the objectification of the technical idea is completed, for this idea is nothing but the understanding of the correspondence between the ideal world of technical and scientific thought and the real world of nature and its technical manipulation. So, if information technology takes this correspondence as its principal perspective, it is precisely the effort to realise the idea itself as a technical artefact, i.e. in the form of the idea itself. So this form of technology cannot transcend itself towards another form of technology. If it transcends itself, it will be towards *something other* than technology. Hegel, in his *Jenaer Realphilosophie*, makes technology transcend itself into the relationship of man and woman. In fact some AI-fanatics, by stating the aim of the discipline as 'making a person'[70] transcend technology towards intersubjectivity. For what technical use is there in making a person?

Against Hollak's position, one may argue that the element of 'power over nature' has not been reflected and included completely in the perspective of information technology. This would require a realisation of power *over this power* inside the technological perspective. This could involve a further self-reflection of technological thought, which could produce a technology of the use of technology, probably not by producing a new kind of *hardware artefact*, but by discovering a new principle of software production enabling mankind effectively to control all use of technology. One may also think of environment problems, requiring such a control as the only way out. But philosophy cannot predict the future, not even its own future. Future philosophies, carrying on such reflections as have led to Hollak's conclusions, will discover whether he was right or wrong in drawing these conclusions, and what this means in a more precise sense than we can now imagine.

Here, however, we are concerned with mathematical thinking. What role does it play in information technology? I have said that it is present there in the form of self-reflection, just as it is in mathematical logic. We have seen that this results in revealing the nature of the mathematical perspective itself. In mathematical logic, this happened in a theoretical way in the discovery of the incompleteness phenomenon. In information technology this happens in a practical way. This technology is based on the use of a general correspondence between a physical process and meaningful thought. What happens in a computer is linked semantically to human intentions of meaning, and therefore the input and output have meaning for us. The machine is working as if it understood and executed our commands. And of course by its design it is made to work that way. Just as a lock and key is made in such a way that one is admitted to a room if one possesses the key. And this is not caused by the physical properties of the key as such, but by its *structure*. The structure is what unites the ideal and the real in this perspective. The ideal structure of thought is reflected in the real structure of a physical process. But we have also seen that real structure is the effect of structuring, which is not

---

70   "The ultimate goal of AI-research (Which we are far from achieving) is to build a person, or, more humbly, an animal." E. Cherniak and D. McDermott, *An Introduction to Artificial Intelligence*, 1985.

only caused by shaping a natural state or process, but also by interpreting this shape as representing a certain structure. Drawing a triangle has only a mathematical meaning if the drawing is understood to represent an ideal triangle. Using a computer for word-processing presupposes that what is seen on the screen is understood as written language. But this, of course, is an obvious presupposition, for we use the signs we have always used - conventionally - in written language. In other words, we endeavour to shape the information processing in such a way that its interpretation may be made to correspond to the structures of our conventional methods of expressing thoughts. Therefore natural language processing is a central issue of AI. But hereby it is revealed that the whole enterprise, not only of information technology, but of all technology, is based on *structural correspondence*. I have said that all information technology is artificial intelligence. Now we have to amend this statement. It only holds in so far as structural correspondence goes. But *how far does structural correspondence go?* Precisely so far as reality is structurability. And that is the original question of this investigation. The limits of information technology are identical to the limits of artificial intelligence, and these are identical to the limits of mathematical thinking. And these limits are determined by the degree of structurability of the world of our experience.

1.7 The limits of mathematical thinking

According to the conception of mathematical thinking given in section 1.4, structurability cannot cover a real material world completely. It is bound to the perspective of the *ideal* opposition of the principles of determination and determinability. Structuring the world is dealing with it *as if* it consisted of ideal - intelligible - matter, receptive of purely structural forms. That we can in fact deal with our experienced world in this way with a considerable amount of success does not prove that mathematical thinking is the only, or even the best recipe for such success. It only proves that there must be some characteristic of the reality underlying the world of experience, that explains this success. In 1.4 I proposed the old Aristotelian relationship of the principles of matter and form - in a 'modernised' interpretation, however - as a possible description of such a characteristic. On that basis we shall now try to answer the question: *what is not reducible to structurability in this world?* In other words: which reality does not depend on the *ideal* sense of the opposition of matter and form *only*.

The most obvious answer to this question seems to be: *life*, and it is not for nothing that Aristotle based his philosophy on the paradigm of life in order to counter the mathematistic tendencies in Platonism. Yet in our time a mathematical biology exists, and recently the mathematical theories of chaos and catastrophe have strengthened the instrumentation enabling the mathematical reconstruction of organic life. Can we still hold the traditional opinion then, first formulated by Hegel and affirmed by Bergson, that if we try to understand life scientifically, the light of our understanding extinguishes at this point? We realise of course, that the phenomenon of life functions only as a test case here. If mathematical thinking is in a certain respect inadequate for understanding life, it is proved that it is in that respect inadequate for understanding physical reality as such, and it will be more than probable then that it will be equally (or more) inadequate for understanding the human spirit. In the terminology of 1.4: If the *real* relationship of matter and form is not reducible to the ideal one, this will manifest itself in all material reality.

Let me first state that I have the highest admiration for the intellectual acuteness and mathematical elegance of the theories of mathematical dynamics on which the attempts towards a mathematical understanding of the phenomena of life are based. I also agree with those[71] who consider the mathematical approach as more adequate than the current mechanistic and vitalistic approaches, and who see in it the possibility of showing that the conflict between the two is futile.

For our philosophical purpose of testing the adequacy of mathematical thinking with respect to the phenomenon of life, we may neglect all questions concerning the empirical adequacy of mathematical biology. Therefore we assume - counterfactually - that this science has completely succeeded in reaching its aim, i.e., it is able to understand the phenomena of life as the realisation of a trajectory in a dynamic structure.[72] The 'formula' of this structure should be determined only by the laws of physics, formulated in some stage of its development; and the trajectory representing, say, the development of a system containing a sufficiently thick layer of the surface of the earth in sufficient detail to contain the process of the evolution of earthly life in all its forms, should start with the representation of an initial condition in a stage in which there was nothing but lifeless matter. Of course, the dynamical system

---

71 "Thus, the 'vitalist' point of view and the 'reductionist' point of view are not at all incompatible (and of the two points of view, contrary to appearance, it is the reductionist point of view which is 'metaphysical', because it requires a reduction to Physico-chemistry which is not established experimentally). René Thom, *Mathematical Models of Morphogenesis*, 1983 p. 23

72 A dynamical structure is a mathematical representation of what can happen in some law-governed system with a finite number of parameters, the parameters of spacial extension and time included. A trajectory is a line in this representation, describing a possible 'history of change' of this system, as determined by its laws. For a very elegant mathematical description of dynamic systems see o.c. note 73

without doubt being highly chaotic, this does not imply that any phenomenon of life be predictable, but it does imply that such a phenomenon, as soon as it appears, can be recognized as represented by a possible continuation of the trajectory.

If we think of such a possibility, we may come to realise that, whatever such a gigantic, almost superhuman theory might be, *it would not be a description of the phenomenon of life!* In order to recognise certain subsystems of the enormous system depicted by the theory as 'living organisms' or types of living organisms and others as 'non-living,' we have to invoke our experience of the world in common sense terms, by which we recognize certain subsystems as e.g. 'this dog' or 'the dinosaur,' so that by our common understanding we know that they belong to the world of living organisms. Certainly the mathematical theory will not 'know' this, unless we bring it in as extra information, e.g. by the terminology we use. Now the whole question becomes a matter of dividing appearance and reality. Is this mathematical structure, describing a dynamic system, 'reality' and our common sense world 'appearance', just as for some cognitive psychologists our experience of perception is appearance and the computing-process in our brain reality? Why is it not the other way round? Or perhaps *both* are appearances, and reality is something entirely different.

In this situation we have the choice of either leaving the question undecided and regarding it as unsolvable, or admitting the possibility of a *third* level of knowledge, on which we can try to mediate between mathematical thinking and common sense. As the first possibility amounts to a refusal to explore all ways open to thought, I consider it to be dogmatic and not worthy of philosophy. So no other course is open than that of distinguishing at least three levels of reflection or conceptuality: 1. The common sense, or *empirical* level; 2. the mathematical level, described in this chapter; and 3. A philosophical level, on which at least the relationship of the other two can be investigated. This is worked out in the following chapter.

Chapter 2
Degrees of reflection: statics

The traditional doctrine of the three degrees of abstraction can already be found in Plato's writings.[73] In present day's terms it can be described as distinguishing three levels of reflection: empirical, mathematical, and metaphysical. These levels give rise to kinds of knowledge, which can be ordered from less to more adequate, which does not mean that the 'lower' kinds are reducible to the 'higher.' How the relationship between the degrees is to be understood, is a philosophical problem in itself, with which I shall deal in Chapter 4. The doctrine of the degrees of abstraction provides starting point in the philosophical tradition, which can shed some light on the problem of mathematism.

Therefore in this chapter I shall explain - and possibly extend - the doctrine, and connect it with more recent epistemological questions concerning mathematical knowledge and its use. Of course, in reality, knowledge cannot be divided into three kinds, according to these degrees. All real knowledge has to do with all three of the levels of reflection in a certain way. In other words: these levels are related in such a way as to give rise to a conceptual motion between them. In this chapter I restrict myself to the 'statics' of the matter. In the next chapter some examples of the 'dynamics' will be elaborated.

2.1 Language, intelligibility and conceptuality

Language is probably as old as mankind. We seem to know what language is. At present you are reading sentences, which means that we are already inside the sphere of language. We use it and understand it. We cannot help using language when we want to articulate thoughts. With every word we say or even think on the nature of language, we affirm by practice that we know the answers already implicitly. Is this a *game*? Ah, another word, do we *mean* anything at all when we say it?

Some philosophers seem to hold that the human spirit is imprisoned in language. But perhaps we imprison *ourselves* in language. Is there no way out? If language is really all we can ever think about, we are not imprisoned, if it is not all there is no need to imprison ourselves. In fact, we could not remain 'inside' the sphere of language, even if we wanted. For the inside of this sphere consists wholly of references to its outside. If language is not considered

---

[73] Especially in the *republic*

to be *about* something, it cannot even be about itself, and we are no longer able to say that language is not about something. It is impossible to deny the intentional character of language for the simple reason that this character is presupposed in the act of denying itself. The use of language consists of saying something about something, either in the form of an assertion or in any other 'illocutionary' form. For in assertion too, something is *done* with language viz. asserting,[74] and in commands, promises etc. something is also asserted: *that* a command is given, and which one; *that* a promise is made, and which one.

If it is thought that language is *about* something, a distinction is made between the act of using language, and what it is about. Now it is perfectly clear that this distinction is not a separability. One could not even think of 'that which language is about,' without using language, even if one is *only* thinking, and one could not really *use* language if it is not about something. Of course the structures of language can be considered as objects in their own right, but just in that respect they are not used. They are also something language can be 'about.' In fact language can be about anything that 'is,' it functions as a pointer towards 'being,' and as far as it *actually functions* this way, it is not itself *considered* as being, but used intentionally.

Distinctions which cannot become separations are very common in philosophical thought: whole/part; matter/form; subject/object; unity/plurality, etc. Yet they seem hard to put up with. There is a tendency either to separate them into dualism or unite them into monism. Either the distinction is denied or the separability is affirmed. Therefore in philosophy we find a never ending oscillation between monisms and dualisms, although at the same time most philosophers seem to be aware of the problem. However painstakingly you try to avoid it, at *some* level it is bound to sneak in again unnoticed.

This also is the fate of the problem of language and being, which is in a certain sense the same as the problem of the intelligibility of the world. For intelligibility of something means that 'the intellect' has access to it, and the only way in which this can manifest itself is that something making any sense

---

74   Frege already discovered this fact when he found that it was impossible to interpret his 'Urteilsstrich' as a function. For the value of a function is an object, whereas the result of asserting cannot be an object. In terms of computer science one could say: it is not a function resulting in a *value*, but a procedure resulting in an *action*. And the action in this case is not again the application of some operation but a confirmation of a conviction.

could be said about it. Now the Sophists[75] were early monists in this respect. They held that it was unnecessary to make sense in order to convince the audience. Of course this was naive, for this opinion was considered to make sense, and even to be *true*, i.e. in accordance with actual practical needs. Plato opposed to the sophists by stating explicitly for the first time the intelligibility of the world by considering the contents of concepts to be realities in their own right. This conception may well have been inspired by the nature of geometrical objects, to which Plato often refers, and which is considered by historians to have been an important factor in the development of his philosophy.[76] Anyway it is a curious fact that for Plato the first paradigm of something beyond language is not the material world, but the immaterial mathematical world. Yet this is understandable. Language as a means of political persuasion, as it was conceived by the sophists belongs to the sphere of *opinion*, which is unstable. In order to reach a region of stable 'context-free' meanings, we have to transcend this sphere. Transcending this sphere takes more than the pragmatics of language in which meanings shift like the waves of the sea. To find at least some constants as a basis of interpretation relative to the context is still the central problem of semantics of natural languages. Semanticists nowadays still cannot think of another solution to this problem than conceiving of more and more sophisticated mathematical representations of meaning.[77] This may be adequate for application in automatic language processing, but it leaves open the question by what we measure this adequacy.

Yet already Plato was aware of a distinction in the *kind* of ideality involved in the use of concepts. The Ideas are supposed to transcend the sphere of mathematical objects. They are 'one of a kind' in contrast to mathematical objects, which have a quasi-substantial individuality, and therefore are the paramount paradigm of numeric plurality.[78] Mathematical objects are also

---

75   At least as far as Plato's caricature of them goes.

76   See I. Toth, Non-Euclidean geometry before Euclid, 1969, See also Plato, Rep.521 ff.

77   Kripkean 'possible worlds' are nothing but mathematical entities. Semantics of natural languages seems unable to do without their kind. The Flemish literary hero Tijl Uylenspiegel inscribed a cross in the edge of the boat, in order to mark the place where the valuable bell of his village was sunk in the lake in order to hide it from the enemy. We sail in the boat of natural language, which of course contains mathematical language, say as its edge. Our mathematical models, therefore, are as dependent on natural language as any other description of meaning. Yet, the domain of natural language is structured by them in an interesting, and perhaps useful way. Only the claim to have solved with such models the mystery of meaning is fallacious.

78   See Republic 521 ff.

recognized to be distinct from material things, and their properties, which we know by sense-perception. From this distinction of three levels of objectivity, the scholastic distinction of three degrees of abstraction was developed, first by Aristotle,[79] then by medieval Arabic and European philosophers. I shall not dwell on historical details or discuss the merits of different - old and modern- representations of this doctrine. Instead I shall eventually gauge my own brand of it, in order to apply it to the problem of mathematics and metaphysics. Nevertheless, it is useful to have a summary of the traditional doctrine at hand, which is as accurate as possible on the basis of my present knowledge of it.

2.2 The traditional doctrine of the degrees of abstraction

In the scholastic tradition the doctrine of the degrees of abstraction was meant to divide the *speculative* sciences - physics, mathematics, metaphysics -, that is, the sciences investigating beings as they are in themselves. The speculative sciences are distinguished from the practical sciences - e.g. logic, ethics -, which aimed at investigating the results of human action in the perspective of their adequacy with respect to the aims of such action. Both speculative and practical sciences were distinguished, as being knowledge from intuition, from knowledge by religious revelation on the one hand, and knowledge by reasoning on the other. It may seem strange that the latter is not included in the speculative and practical sciences, but one can of course draw conclusions from a mixed set of premises, as is nowadays actually done in many disciplines, such as 'management science'; 'computer science'; 'communication science' etcetera. The scholastic distinction of the sciences is therefore based exclusively on a distinction of the origins of their forms of *knowledge*, which enables us to consider it more widely, as producing a division of all human knowledge, whether it has the explicit form of a specialized discipline or not.

The degrees of abstraction are traditionally distinguished by what they abstract *from*. The first degree abstracts from the factual *individual situation* in which a concept is formed or used. The result of this abstraction consists of general, empirical concepts. The second degree abstracts from 'sensible matter,' i.e. from every content, the meaning of which cannot be understood without referring to actual perception. *Change* and *quality* are understood as the main characteristics of such contents. The result consists of mathematical

---

[79] See Aristotle, Metaph. E, 1, 1026a13-19 and K, 7, 1064b1-6.

concepts, i.e. concepts of structurability. The third degree is supposed to abstract from the *relativity* with respect to the acts of perceiving and structuring presupposed in the other two. It does not abstract from a positive content, but only from a *form* in which the contents of experience are understood.

The first degree implies relativity with respect to the act of sense-perception, the second implies relativity with respect to the act of structuring. Abstraction from these relativities must result in notions determining the object *in itself*, and not merely with respect to certain acts or attitudes of the subject only. So here, what we are supposed to abstract from, is itself abstraction, in the sense of empirical generalisation or mathematical structuring. Therefore it is sometimes said that abstraction of the third degree is abstraction from abstraction.

Of course especially the third degree of abstraction asks for sceptical criticism. How can we know 'the object in itself'? No one less than Immanuel Kant has warned us against the illusion that such a thing is possible. The problem for the critics of the third degree of abstraction however is, that a gradation of *knowledge* is connected to the gradation of abstraction. If the third degree appears to be impossible, the other two must also fall. For if we cannot know any *object* in itself, we cannot know any *relation* in itself either, for the word 'object' cannot be used here in any other sense yet than 'something knowable.' A meaningful distinction of 'object' and 'relation' as ontic categories would obviously presuppose knowledge of the rejected 'absolute' kind. So the logical coherence of the doctrine of the degrees of abstraction gives it a 'take it or leave it' character. It is not divisible by Occam's razor. Therefore probably, modern philosophy has left it completely, attacking *directly* the problem of the possibility of the kind of knowledge involved in the paradigm of the modern era: the natural sciences. Moreover the idea of *degrees* of knowledge seems to fit better in the hierarchical world-picture of the middle ages, than in the modern democracy of knowledge, in which the 'best distributed good in all the world,' i.e. common sense,[80] functions as a universal and impartial judge, provided it is led by scientific method.

---

80   "Common sense is the best distributed good in the world: for everyone supposes he is so well endowed with it, that even those who are satisfied hardest in every other matter, do not habitually desire more of it than they possess." This is the opening sentence of René Descartes' *Discours de la méthode*. In fact it is *mathematical* thinking, which will be established as the ultimate judge of science. One glance at Descartes *Regulae* can affirm this. This thinking does not allow of an epistemological nor of an ontological gradation. It is based on a 'grey' ontology, as Husserl formulated it. The object of any knowledge is *res extensa*, not understood in its concrete material and spacial sense, but in an abstract mathematical sense, in which it can be grasped by 'clear and distinct' ideas.

Now in what respect are the 'higher' degrees of abstraction supposed to give more adequate knowledge than the 'lower'? Knowing the tradition of Aristotelian thinking, one is not surprised to hear the answer: *as means of access to knowledge*. A distinction of degrees would be rather superficial if the differences were merely accidental. They must of course be essential, which means that they must refer to the essence of knowledge. Now one may have different views of many dimensions of the problems of epistemology, but any *discussion* of these problems presupposes an idea of what it is all about. Now a description of this minimal agreement on the nature of knowledge may be that *what is known* has to be really present in *that of which it is known*. I can only know grass to be green, if it really is green, otherwise I do probably guess it, but I do not *know* it. That is to say that there must be a relationship of *truth* between what is known of an object, and the object of which it is said to be known. Now the degrees of abstraction clearly differ in this. The green, actually is present in the grass only in so far as the grass is *perceived* as being green. Knowledge given by the first degree of abstraction does not concern something as it is in itself, but only as it is with respect to our perception. Also in mathematical knowledge there is such a relativity, but it is somewhat less severe. A Euclidean triangle has an angle-sum of two right angles in so far as it is *structured* as a Euclidean triangle, i.e. in so far as I have *postulated* that it is such a triangle. But unlike the grass, the triangle has no other existence than being the result of a relationship to a certain act: the act of structuring. So knowledge given by the second degree does not concern something as it is in itself either, but only with respect to our structuring. Perception is not supposed to *create* its object, structuring in fact *does* so,[81] therefore the *result* of structuring is indeed known as it is in itself. On the other hand, we have to confess that the triangle is less *real* than the grass. So what we win in adequacy, is lost in reality. In pure mathematics we can have perfectly adequate knowledge, but in mathematical sciences of the perceivable world, knowledge is relative with respect to our model.

Of course the third degree is supposed to characterise knowledge of a non-relative kind. The only relationship involved here is the relationship of

---

81 I am not occupying a Platonic position with respect to the existence of mathematical objects here, but, in accordance with the subject of the degrees of abstraction rather an Aristotelian position. It can also be noticed that traditional *empiricism*, by identifying the *content* of perception with the *act* of perception, is really a form of mathematism in so far as it does not recognize an independent object of perception.

knowing, and this is by definition just the relationship in which the known enters as *itself*, as identical within or without the relation.[82]

By denying the possibility of this degree of knowledge, we are obliged to deny the other degrees too. So a sceptical attitude has the choice of either rejecting the whole idea, or asking whether by gaining perfect adequacy on the third level, we also necessarily loose something. This question will be discussed extensively in 2.5, for it is one of the central issues of this book.

This section on the traditional doctrine must be concluded with a remark concerning the scholastic distinction between the 'first' and 'second' intention. This is most obvious in mathematical thought: one can consider a perceivable thing or process in the perspective of its structurability by actually understanding it in accordance with a certain definite structure, e.g. the space of bodily motion considered as Euclidean space. This is the 'first,' or 'direct' intention. Secondly, however, we can be aware of the ideal nature of Euclidean geometry, and realize that what we have actually done is *applying* this geometry to the space of our bodily motion. This is the second or oblique intention, in which we are aware of the *relationship* of the result of abstraction with respect to its origin in experience.

In the first degree of abstraction the first intention is nothing but the unquestioned conceptualization of the world of experience; in the second intention we become aware of the fact that we continuously apply abstract concepts.

For the third degree the distinction is between the fundamental perspectives in which we tacitly understand our world and explicit philosophical reflection.

2.3 Abstraction and reflection

Abstraction in itself does not produce knowledge in the full sense of the word. The conceptual abstraction of the first degree produces a conceptualisation of the world of experience, but empirical knowledge of this world is expressed in *judgements*, in which the concepts are used in order to determine a state of affairs. Judgement can be understood as a form of *reflection*[83]

---

82 "Nur innerhalb des Begriffs gibt es etwas, das ausserhalb des Begriffs existiert", Bruno Liebrucks, *Sprache und Bewußtsein* I, Vorrede 1964-1969

83 I use the term 'reflection' here in the literal sense of 'bending backwards,' not in the sense of 'giving account explicitly.' This last sense has much to do with the scholastic distinction between primary and secondary intention. In my terminology reflection can be implicit as well as explicit, and both are to be distinguished from *self-reflection*, which considers

in which the applicability of a concept to a real experience is understood and expressed. According to classical logic a concept cannot be true or false in itself. It is formed by abstractive intuition, inspired by experience, but it does not contain the pretention really to *apply* to this or any other particular experience. In a judgement about real experience, the concept used as a *subject of judgement* is not actually applied to anything. It only serves to *point* towards the experienced state of affairs the judgement is *about*; for an insider it can be replaced by 'it' or 'you know what I mean.' It is the *predicate-concept* which is really applied to this state of affairs, which is thereby identified with the present content of this concept.[84] So conceptual abstraction is only one side of empirical knowledge, which is completed by judgement in order to become a motion of reflection from, and again towards experience. This motion is what we usually call 'thinking,' and it is involved in all our theoretical and practical judgements in ordinary life. It constitutes the 'conceptualized world' which is the most obvious form of reality for us.

Reflection, however, generates self-consciousness. A judgement evokes the question: why? Each man is mortal. Why? Because each man is a living being, and each living being is mortal. As an answer to the question 'why?' a new concept is interpolated between the subject- and predicate-concept. This new concept functions as an explanation of the judgement, which has thereby become a conclusion. The classical doctrine of syllogism is a reconstruction of this motion of thought, generated by the self-reflective tendency of the reflective motion of judgement. One may notice that this self-reflection returns to the realm of *concepts* again. It can result, as Aristotle realized in his posterior analytics, in a *theory*, which essentially is a conceptual system. Explicit knowledge tends to take the systematic shape of a conceptual ordering, and this ordering can be in its turn subject to 'second order' knowledge, which is in this case called 'logical' knowledge.

So the first degree of reflection produces three kinds of knowledge: 1. Common sense knowledge, expressible in judgements. 2. Systematized empirical knowledge, expressed in empirical theories. 3. Logical knowledge about the adequacy of the judgements and theories mentioned in 1 and 2.

---

the contents of a mental act in its own ideal form. So, the continuous judgements about our world performed throughout our lives are examples of implicit reflection, which becomes explicit as soon as we are going to ask epistemological questions about such judgements. If we only consider them in their own nature, *as* judgements, we consider them in the form of self-reflection, which means that we are doing logic.

84   Cf. Henry Veatch, *Realism and Nominalism Revisited*, 1954.

In mathematical reflection, the same motions can be distinguished. Mathematical abstraction does not yet produce knowledge in the full sense of the word. It results in the insight that a specific kind of structure is possible. Such a structure in itself cannot be true or false. In order to be so, it must be *applied* to some experienced reality, which thereby becomes *structured* according to this structure. Structuring the world of experience, traditionally understood as *measurement*, here takes the place of judgement in empirical knowledge. But this motion of reflection generates a kind of self-reflection too. One may ask why this structure applies, and in order to explain this, other structures are interpolated, which asks for reasoning about the structures in themselves, and for a systematic investigation of them. The realm of theory, proceeding from this self-reflective form of mathematical reflection is developed in what is called pure mathematics. It involves an ordering of structures as such, and an investigation of their relationships. Because mathematical reflection 'substantialises' its results, self-reflection can take one step more than it can do in classical logic. It can investigate its *own* structure, which is realized in mathematical logic.

The second degree of reflection also produces three kinds of knowledge: 1. Knowledge of the structurability of the world of experience, expressed in theoretical or practical structuring, that means: scientific models and technical designs. 2. Pure mathematical knowledge. 3. Knowledge of the structurability of mathematical reasoning itself: metamathematics or mathematical logic.

The third degree of reflection can be understood in somewhat the same way. Insight into principles, however, must be understood as already constituting knowledge in the full sense. For in so far as they really open up a meaningful perspective, they are necessarily *real* principles, and there is no distinction in content between the principles as they are understood, and the principles as they are in reality. This is the insight implicitly involved in ordinary life as well as in science. It makes us understand experience in a certain perspective and think about it consistently in that perspective. In this way, our insight into principles returns to immediate experience, already before we have become conscious of it.

But insight as a subjective act, in so far as it is still purely intuitive is not yet insight into the full sense. Only if it has produced the ability to *explain* by reference to experience that these principles really *are* principles of the experienced world, the insight has become concrete and real itself. This may be done by analysis and description of the experienced world in the perspective originating in those principles, thereby interpreting the results of scientific investigation and common sense experience, as it is done for instance in classical and modern metaphysics, and also in analytical and phenomenological philosophy. Just like judgement and structuring, such philosophizing raises many questions. The attempt to answer them, again leads to the tendency to systematize. But here the self-reflection is immediate. The philosopher cannot go about his business while leaving the reflection about it to someone else, while the mathematician can peacefully leave the reflection about the structure of his reasoning to the mathematical logician. In philosophy the self-reflection has to be built into the activity of philosophizing itself. This self-reflective tendency is present in all philosophy, but it may be said to have culminated in Hegel's speculative dialectics, in which, as on the mathematical level, a second degree of self-reflection is present in the form of the Science of Logic. However, as logic cannot replace reasoning, and mathematics cannot replace the mathematical sciences, and metamathematics cannot replace mathematics, so dialectical rigour cannot replace speculative insight. There must be a certain equilibrium between the two. After Hegel, this equilibrium has been disturbed by a wave of new insights and perspectives caused by the development of the empirical and mathematical sciences. That is the probable reason why nowadays philosophy seems to have acquired an anti-systematic tendency. The awareness that this is an exaggeration of the same order as the over-systematisation criticized by it, belongs to an even more sophisticated dimension of philosophical self-reflection. It has to do with the self-insight into the principally limited capacities of the human mind. Intuition and objectification can neither be separated nor be identified. In whatever philosophical terminology we express this state of affairs, we always have to accept it as belonging to the human condition. We can make a separate discipline for questioning it, for instance philosophical anthropology, but that does not stop it pervading all our philosophizing. Therefore philosophical thought, unlike mathematical thought, cannot place itself safely outside its subject-matter. It can never be 'theoretical' in the sense of neutrally observing or exploring its field of interest, for it knows that its methods, just in so far as they are methods might deflect it from its goal, and on the other hand its intuition is not powerful enough to reach this goal directly. Yet the *ideal* of pure explicit knowledge of principles, which is metaphysics, remains the motivating force of all philosophizing.

The third degree of reflection again produces three kinds of knowledge. 1. Effective implicit insight into principles making our concepts and structures meaningful with respect to the world of experience. This is the most important and the most robust kind of knowledge. Philosophers may doubt about the true nature of arithmetic, for instance, but nobody doubts the real possibility of counting and calculating. 2. Philosophical knowledge, resulting from the attempt to express this implicit knowledge in an explicit form. 3. Metaphysical knowledge as the ideal of pure explicit knowledge of principles.

2.4 Empirical reflection

Our experience of the world is highly determined by the culture we live in. What we perceive, depends on the concepts we possess to describe it, or even think of it. Cultural anthropology has convinced us that cultural conceptual networks are difficult to compare with each other. What seems crazy, chaotic or superstitious to us, may be completely logical from the point of view of another culture.

Yet, on the other hand, cultures influence each other. All cultures in the world have somehow integrated many elements of western culture now, but western culture in its turn has integrated substantial parts of other cultures in its long history. Cultural pluriformity too, presupposes some basis of understanding between the partial cultures involved in it. The plurality of cultures is not mathematically discrete.

This situation corresponds to the character of empirical knowledge. What counts as, and can be acquired as empirical knowledge very strongly depends on a conceptual apparatus, a language in which we think.[85] Language is more than a medium of communication. It expresses the structure of our life-world. On the other hand, a conceptual apparatus contained in a specific language is by no means completely arbitrary, and conceptual systems never are so different that they are not in a certain measure intertranslatable. The same holds for those elements of the conceptualisation of experience which are not obviously connected with linguistic expressions. They constitute something like Lacan's 'symbolic order,' which governs our unconscious as well as our

---

85 "Language is not merely a more or less systematic inventory of the various items of experience which seem relevant to the individual, as is often naively assumed, but it is also a self-contained, creative symbolic organisation, which not only refers to experience largely acquired without its help but actually defines experience for us by reason of its formal completeness, and because of our unconscious projection of its implicit expectations into the field of experience. In this respect language is very much like a mathematical system which, also, records experience in the truest sense of the word, only in its crudest beginnings, but, as time goes on, becomes elaborated into a self-contained conceptual system which previsages all possible experience in accordance with certain accepted formal limitations." Edward Sapir, Conceptual categories in primitive languages, 1931, p. 578. This idea is usually called the Sapir-Whorf Hypothesis. Cf. Harry Hoyer, *Language in Culture*, 1954

conscious life. It may be culturally determined, but it is never arbitrary. "It's madness, but there is a method in the madness."

This is the way concepts function in life. They are *meaningful distinctions* in the world of experience, connected with language and culture, but also with something else, which we may call 'nature,' or better, the nature of what we are talking and thinking and feeling about, and which is not obtainable in pure, 'unspoken' form. In practical life we use the conceptual schemes of language with perfect faith in their capacity to express adequately the true nature of the contents of the experienced world. Even if we admit, that other conceptualisations are possible, we do not lose faith in our own, for we experience them as adequate and meaningful.

Consider e.g. the distinction of substantives and adjectives. Aristotle is nowadays sometimes thought to have been naive in assuming that there really are substances and -accidental or necessary- properties. The 'built-in metaphysics' of indo-european languages may not be universally accepted in other cultures. This could be the case, but it does not alter the fact that the distinction is *meaningful*. One cannot divide experience arbitrarily, as it is done in Luis Borges fantasy of 'a certain chinese encyclopedia,' in which the animals are divided into:

> "a) Those belonging to the Emperor; b) embalmed ones; c) tame ones; d) sucking-pigs; e) sirens; f) fable-animals; g) free running dogs; h) those mentioned in this division; i) those beating around like madmen; j) innumerable ones; k) those drawn with a fine brush of camel hair; l) etcetera; m) those who just broke a jar; n) those looking like a fly from a great distance"

This list might nowadays be the result of some crazy computer-processing of a meaningful text, and of course it is not really a division. But who guarantees that our common language is based on proper divisions? And even Borges' fantasy is expressed in meaningful language, otherwise we could not even laugh about it. Edward Sapir[86] compares languages to mathematical systems, in fact *originating* in experience, but then becoming closed systems, structuring further experience according to their formal rules. But such mathematical systems too, presuppose a meaningful context, and they are usually expressions of the discovery of the possibility of a certain structure. Moreover such a point of view already involves a reflection of the second degree. It is perfectly possible to look at the empirical level from a

---

86   See note 86

mathematical point of view, and usually when this is done, it is claimed that the mathematical perspective teaches us the 'true state of affairs'. But this, of course, is precisely what I called 'mathematism'.

The first degree of reflection can be said to result in a meaningful division of the world of experience, by means of which we are able to express knowledge about this world. This knowledge primarily concerns the world *as we experience it* given the cultural situation in which we live. It is expressed in all our empirical judgements, understood by those who share our cultural situation sufficiently. This knowledge belongs to what Husserl calls 'natural consciousness,' and expresses the fabric of our life-world.

It has often been tried to reduce the data of our natural consciousness to sheer appearance. Yet, I claim that it is *knowledge*! In what respect can this claim be made, and how can it be philosophically justified? We have seen that the hypothesis that our division of the world of experience is completely arbitrary is absurd. All cultural relativism is based on the concept of a culture as a meaningful totality of human life. This totality may very well imply a metaphysical position, but it is certainly not *reducible* to the arbitrary choice of such a position. It characterises a *real* possibility of meaningful human life, and this possibility is *known* in exercising it. The world as we experience it is not an illusion, but a form of *reality*. This reality is not only a source of inspiration for the discovery of a certain structure, as it is in the mathematical case, but it remains involved in all actions. We do not build a world of pure, imaginary structures, but we live in a conceptualised world. And this world, as it is pervaded by language and conceptualisations relative to our perceptions, is *real*. It does not consist of 'pure phenomena' or of an abstract 'language-game' or 'discourse' or 'text' or 'symbolic order.' All these terms may be adequate, provided it is made clear that they do not refer to a bloodless 'system,' but to the reality of our lives, that is what makes the difference. Of this reality we do indeed have knowledge in the genuine sense. We may say that 'the grass is green' because that in fact is the case in the world *we live in*. No cultural or linguistic relativism can alter this fact, for the relation between content and form of experience, on which the relativism is based, *is* precisely this reality. The problem with most relativism in fact is, that it does not believe in itself. It says e.g. 'the world is structured by our words' but silently it is believed that the world only *seems* to be so structured, that its relativity with respect to the form of experience is only apparent. But what is empirically known is ontologically relative and not merely epistemologically. Therefore it can be adequately known by means of our conceptualisation.

But what with different cultures? Do they have different realities then, and how can they communicate? Here we have to be on guard for unrecognized abstractions. The difficulties of intercultural communication are real facts of life, not theoretical constructs. The worlds of different cultures clearly have something in common, in spite of their structural differences. For completely ideal, mathematical structures, this would be difficult to understand, for these do only have in common, what is *postulated* to be common to them. But no human instance can postulate community between cultures. This community must be *realised* by people belonging to these cultures themselves. That this is possible, can be seen as another proof of the reality of the cultural worlds, in distinction from the ideality of mathematical 'worlds.'

What does it mean to say that our living-world is 'real'? In the first place, this word is used in order to distinguish this world from the 'ideal' worlds of mathematical construction, belonging to the second degree of reflection. But what then distinguishes these two? The classical answer is of course, that the first is perceptible by our senses, and the second is not. The concepts of the first degree of reflection are only meaningful in connection with sense-perception. Our daily language is full of images and metaphors referring to perceivable qualities. Many of the more abstract concepts are related to our activities of orientation in space and time. Our life-world is a world of sense as much as it is a world of words and meanings. Yet there is a doubt concerning the reality of this world, which is as classical as the answer given above. Do not our senses often deceive us? Can it not be that they always do? Possibly, but what is the standard by which we measure deception? If it is the world of sense itself, we affirm its trustworthiness by every unmasking of a deception. If it is the intellect, as it was for Parmenides and his followers, it is the whole sphere of empirical knowledge which is called in question. This may take place either on the mathematical or on the metaphysical level. On the mathematical level there is much reason to doubt the reality of the world of sense. For what is real for mathematical thinking is found by postulation and proof, and these are rarely trustworthy in practical life. From a mathematical point of view, this is so much the worse for practical life. On the other hand, practical life is strong enough to defend itself against this kind of mathematism, and it will not lose its self-confidence because of it.[87] Yet the attitude of trying to substitute a mathematical model for practical and perceptible reality seems to be inexterminable nowadays. Therefore, metaphysics has to be the judge in this conflict, and it will have a difficult job

---

87 Jonathan Swift characterised a person as being "..as dull in drawing inferences from daily life as any dutch professor of mathematics"

there, as we shall see in chapter **5**. There also is the place to clarify further what I mean by the reality of the life-world.

Now it is said, that the life-world is a world of sense as well as a world of words. But what is the relationship between the two, or, why is this world a result of reflection and *abstraction*? The original idea was, that we abstract from the individual case as such, in order to form a general concept. The result is, that all individual situations are covered by hurdles of concepts, and this conceptual structure appears as its reality. The thing we *call* a tree actually *is* a tree for us, in spite of its individual uniqueness. We still know of this uniqueness, but it is given at the same time as a case of the general concept. If, by a pathological state, we loose the immediacy of this connection, it is experienced as very threatening, for what we loose then, is our life-world. By reflection we still always know, that this life-world, in which all things have names, transcends sense-experience as such. We can hardly imagine a world in which nothing has a name, and in which only what is here and now exists. Yet we know that this is a *side* of our world of experience, with respect to which the conceptual world is abstract. So abstraction is a *relation* here, between two sides of the life-world. On the one hand we live as vital beings in the immediacy of our emotions and sensations, yet being aware of this immediacy; on the other hand we live as cultural beings in the symbolic order of words, yet having emotions and sensations. We cannot separate these sides, neither can we unite them so as to become indistinguishable. This is the human condition, referred to by many names: bodily spirit; excentricity; existence; freedom.[88] In a certain sense, this is what is called 'human practice'[89] as a basic level of our being to which all other levels are relative, even if they intend the absolute. But it is a mistake to understand this level in a monolithic way. It has its own inner relationships and tensions, from which other levels arise. One of these inner relationships is the reflection of the first degree. It is the relation of individual situations to their conceptualisations.

Individual situations or 'moments of life' do not form a disconnected set of single experiences. In 'stream of consciousness' literature, they must be pictured this way, because literature is like cinematography: motion must be suggested by a succession of images. But just as in cinematography, the resulting illusion truly depicts the continuous change of reality. Henri Bergson

---

88   The terms are respectively found in the writings of J.H.A. Hollak, H. Plessner, J.P. Sartre, and G.W.F. Hegel.

89   See S. Stenlund, *Language and Philosophical Problems*, 1991

introduced the concept of *duration* for characterising the nature of this continuous change. It is misleading to think of a mathematical continuum, e.g. a time axis, here. In such a continuum there are point-like instances, which are mutually disjunct, and united only by an abstract topological structure. But the instances of our life are by no means disjunct or point-like. Our whole past is present in our present moment. Not as ordered by a succession of time, but as ordered by the meaningful totality of experience. Remembering something, is separating it from this unity, and connecting it to our notion of a past stretching out backward in time. This notion only comes up in so far as we are reflecting about our lives, not in so far as we are actually living. Abstraction is another form of such a separation of something from the unity of experience. What is separated here is placed in the conceptual order of language. Comparing it to the results of remembering gives us the impression that a concept covers many instances having something in common. But that is only a post-factum description, not necessarily the *origin* of this abstraction[90] For somehow the 'something in common' must have been picked up from experience, and this has nothing to do with the plurality of cases, which only appears *after* a uniting concept is formed. We cannot recognise a second instance as 'the same' as a first instance, if we have not already noticed the common characteristic in this first instance[91] implicitly or explicitly. So conceptual abstraction cannot have a plurality of cases as its starting point, but only a single case. And, properly spoken, not even that, for even in order to recognize it as a single case, the concept must already be formed. It is the *totality of experience* itself, from which abstraction can only depart. The distinction of unity and plurality in this totality is itself a result of this abstraction, for it introduces a distinction within the totality, just as remembering does.

This distinction is common to discovery, learning and use of a concept. For a concept can only be learnt or used by discovering in an actual situation the distinction indicated by it. If a child learns a word it learns how, and to *what* to apply it. This knowledge is precisely knowledge of the concept

---

90 Ernst Cassirer, in his book *Substanzbegriff und Funktionsbegriff* (There exists an English translation, see Literature list) pictures the classical idea of abstraction as this finding of a common characteristic in a plurality of cases. So it becomes a kind of *structuring* this plurality as a *set*. Seen from this mathematical point of view this is only one, rather primitive, kind of structuring, and modern mathematics has to offer much more sophisticated kinds.

91 Aristotle, in his famous text of Posterior Analytics II, 19 (99b15-100b15), is perfectly aware of this. He uses the image of military men, one of whom stops, and faces the enemy during a retreat. Then the others follow his example. Of course here de decisive act is the first man's. Its adequacy is only *affirmed* by the others following him.

connected to the word. And if the word is used, it is supposed to apply to some situation in which its conceptual distinction is present. Our awareness of this distinction is triggered by the word, which is exactly its function. But de Saussure was right, there is no necessary connection of *content* between the word-sound, and this distinction. In this sense the knowledge of the conceptual distinction is *independent* of language. It is introduced into the totality of experience by *abstraction*, i.e. by insight into the possibility of making a meaningful distinction.

At this point we can ask the old[92] epistemological question whether the meaning of the distinction originates in the world or in our mind.

The answer depends on the level of reflection from which we try to give it. It could be argued that ultimately it is a philosophical question, so the philosophical level must be decisive. But are we so certain that philosophers know better than ordinary people? Or, formulated in a less self-defeating way, is it totally irrelevant how such questions would be answered from the level they concern? I think that it is very relevant, especially for philosophy, and that is the truth contained in the idea of 'common language philosophy.'

Now for empirical reflection the world is the conceptualised life-world. If a new concept is formed, the meaning of the distinction it introduces is discovered as present in the world in this sense. And since for empirical reflection the world is not an abstract system but the reality we live in, the concept indicates an objective meaningful distinction.

If we *notice* that the whole life-world is structured by our conceptualisation and is therefore not innocent of the creative activity of the human spirit, we have left the level of empirical reflection. It is because of our activity of structuring then, that concepts have meaning. They indicate vertices in the structural webs we project into experience. This is, how the world of empirical reflection appears from the point of view of mathematical reflection. The difference between these viewpoints is a philosophical controversy only for philosophers taking either empirical or mathematical reflection as a paradigm for philosophical reflection.

---

92   The whole problem is dealt with in a very subtle way already by Nicolas Cusanus, who was aware of the receptive as well as of the creative side of human knowledge. Against the scholastic tradition he considered the activity of the mind in the act of knowing no longer as purely *formal*, as only bringing about the form of intelligibility, but as creating a content of its own, a world of *signs* which the mind tries to assimilate to real being. This assimilation process is never-ending, yet it is meaningful as a continuous and progressive contemplation of the divine. Many of the modern discussions about reality and language are surpassed in subtlety and acuteness by this medieval cardinal. See e.g. Nicolai de Cusa, *De coniecturis* 1972. An excellent comment is: Theo van Velthoven, *Gottesschau und menschliche Kreativität* 1977.

2.4 Mathematical reflection

Making distinctions in the totality of experience is already structuring. But only in a certain perspective, the perspective of separability of what is distinguished. If we call some animals cows and others horses, this distinction becomes structuring as soon as animals of both kinds are together in a meadow, where they each have their own place, separable from the places of the others. Distinction is not the same operation as separation, even in thought. As directions 'upward' and 'downward' are distinct, but they cannot be separated actually or in imagination. This is true for all pairs of converse relations. We can of course *represent* these relations by separate objects, e.g. sets of ordered pairs, as it is done in set-theory. But that does not make them separable themselves. Now the experiential situations distinguished by the concepts we use in daily life are usually either separate as individuals in space and time or can be *thought of* as separated in such a way. The totality of experience can be conceived of as falling apart in a plurality of separate individuals. It can be regarded as being an *external world*, which means that it is external to our subjectivity, but also that it consists of parts which are external to each other. It is Descartes' *res extensa*, which is the world of experience, seen in the perspective of structurability. In such a world, conceptualising is structuring. Now we do not only conceptualize kinds and properties of things, but also relations and other coherences. In the perspective of externality, these give rise to all kinds of structure.[93] Among the oldest concepts noticed to introduce structure are the quantitative ones in the stricter sense of the word: equal; more; less; the natural number concepts (in the concrete sense of: number *of* things); the concepts of measure and figure. They introduce the numerical and figurative structures which play an important role in practical life. Apart from conceptualizing our world, we are also continually structuring it. Technology, of course is a clear sign of it. In order to make the most primitive instrument, already presupposes structural thinking. Any technical design involves such structuring, consciously or unconsciously.

The elementary structures, later called quantities, are investigated in early mathematics. These structures have slowly grown to be regarded as entities worth studying for their own sake. In old Babylonian, Egyptian and Chinese mathematics they are still studied without the explicit awareness that the knowledge thus obtained is of a type quite different from ordinary conceptual knowledge of the experienced world. This awareness arose in Greek mathematical thinking, and gave rise to philosophical amazement. For what is so special about quantity that it makes such 'immaterial' knowledge possible?

---

93   See note 91

In the Platonic Academy it was discovered that it was possible to think of the objects of geometry as of separate entities of an ideal nature. And to think of them in this way is mathematical abstraction, later called the *second degree* of abstraction.

The ancient Greeks, and in fact almost all mathematicians and philosophers till the nineteenth century understood this abstraction as a relation between *real structures*, and *ideal structures* having the same structural characteristics, but different modes of being. In the nineteenth century it was gradually discovered that mathematical thinking could, and in fact already for some ages did, produce structures not corresponding to anything that was beforehand found in experience, e.g. complex numbers and non-Euclidean spaces. But the even more amazing discovery was, that these structures could nevertheless be *applied* to experience. Mathematics could therefore be understood as a creative enterprise of ideal construction, which in its turn could structure the experienced world. Now even the older mathematical disciplines, and in fact all structuring activity in ordinary life, can be seen this way, and this means that mathematical reflection is not only an activity constituting a special science, but it is involved in *all* structuring of the world. We now have to leave the point of view from which this reflection constitutes a relation between completely determined real and ideal structures. Just as reflection of the first degree *creates* concepts, reflection of the second degree creates structures. Both are based on the insight into a *possibility*. Conceptual reflection on the insight into the possibility of meaningful distinctions in the totality of experience, mathematical reflection on the insight into the *structurability* of this totality.

Structurability is a property of reality as we experience it. We can not only *change* things so as to be able to represent a certain structure, but also *interpret* them according to such a structure. Nothing has only one definite structure, for structure is the effect of dividing and separating the resulting parts *in thought*. This has been illustrated impressively by R.M. Pirsig in his book *Zen and the Art of Motorcycle Maintenance*, although his text does not clearly distinguish conceptualization and structuring[94]:

> A motorcycle may be divided for purposes of classical rational analysis by means of its component assemblies and by means of its functions.
> If divided by means of its component assemblies, its most basic division is into a power assembly and a running assembly.

---

[94] This is characteristic for the attitude of mathematism, in which the meaningful embedding of concepts in the totality of a cultural context is overlooked, and only the structuring effect of conceptualization appears.

The power assembly may be divided into the engine and the power delivery system. The engine will be taken up first.
The engine consists of a housing containing a power train, a fuel-air system, an ignition system, a feedback system and a lubrication system
The power train consists of cylinders, pistons, connecting rods, a crankshaft and a flywheel.
The fuel-air system components, which are part of the engine, consist of a gas tank and filter, an air cleaner, a carburettor, valves and exhaust pipes.
The ignition system consists of an alternator, a rectifier, a battery, a high voltage coil and spark plugs.
The feedback system consists of a cam chain, a camshaft, tappets and a distributor.
The lubrication system consists of an oil pump and channels throughout the housing for distribution of the oil............"
....
".......there is a knife moving here. A very deadly one; an intellectual scalpel so swift and so sharp you sometimes don't see it moving. You get the illusion that all those parts are just there and are being named as they exist. But they can be named quite differently and organised quite differently depending on how the knife moves."

Even if we have made things so, that they have a certain structure, still they can be interpreted, and even *used* according to a quite different structure. It is a well known fact that many instruments are apt to be used for tasks for which they were never designed. We open jars of jam with coins, do word-processing on computers, fry eggs using the cheese knife, make music on oil barrels and washboards, make lamps or garden fences out of cartwheels, etc.

Not only technology, but also pure mathematics witnesses to the freedom of structuring. The only limit to the design of new structure-concepts by systems of mathematical axioms seems to be the requirement of consistency. This requirement is sometimes thought of as a sheer formality, which is easy to fulfil. Since Gödel's theorem, we know however, that only *in*consistency is a formality. We cannot prove the consistency of a system, unless we presuppose more knowledge about its subject matter than is made explicit in the system itself. The only warrant for consistency therefore must be this subject matter itself. Consistency is not a formal, but a 'material' property of a theory. A mathematical theory is consistent if and only if it has a mathematical *meaning*, i.e. it is an adequate specification of the principle of structurability. This principle constitutes and limits the freedom of mathematical construction. Investigating the possibilities of such constructions, i.e. performing mathematical research, is investigating the scope of the principle of structurability. As this principle governs what is *possible* in the realm of structure, the only way to investigate it is by *actualising*. In technology this is done in the real world of experience, in mathematics it is done in the ideal world of abstract structure. And this ideal world is based on what was called in chapter 1 'mathematical abstraction in the strict sense,' which means understanding that structurability is a real principle of the experienced world of such a nature that it can constitute a world of its own in our imaginative thought. It is this simple act of genuine understanding that gives mathematics all the glamour of exactness. For this means that an intellectual perspective is found, which is independent of sense perception in its concept-formation, and yet expresses such a fundamental property of the experienced world, that it generates *power* over this world. The ancient Greeks were right when they felt that the possibility of pure geometry was a very great discovery. Here the intellect seemed to be on its own ground, not hampered by the inaccuracy and insecurity of the senses. But the greatest philosophers of that time already saw another consequence. If we are able to understand *one* fundamental principle, perhaps we can understand *more* than one. Does not mathematics by its own nature lead towards metaphysics, which is knowledge of principles as such?

Mathematics is based on the understanding of the principle of structurability, but this understanding remains as *implicit* in it as it is in the structuring activity of ordinary life. It gives mathematics the status of a science, however, for mathematical knowledge is *explicit* knowledge of the *consequences* of this principle, of the particular possibilities of structuring our experience. Explicit knowledge of the principle itself, however, does not belong to mathematics, which therefore cannot express its own foundation. One of the most genial attempts to do so, Cantor's axiom of comprehension, clearly shows where the limit lies. It states that a property of certain mathematical objects gives rise to a *set*, which is itself a mathematical object completely determined by its elements, i.e. those objects possessing the property. In our terminology this amounts to the statement that *any concept structures the world*. For a property corresponds to a conceptual distinction, an expression of empirical knowledge. A set on the other hand corresponds to the *extension* of the concept. So the axiom states that one can abstract from the meaning of the concept, and thereby conceive of the extension as an ideal entity. This is an adequate expression of what happens in mathematical abstraction, and therefore a true foundation of mathematics. But as soon as it is understood as something *within* the realm of mathematics, as an axiom constituting a mathematical theory, it becomes inconsistent.

For the property has to be a *mathematical* property now, derived from the notion of *set* itself. Now this notion is based on the possibility of abstraction from the content of the property forming it. That means, that this property necessarily forms an *outer limit* for it. So a set is by definition something limited from the outside.[95] Its limit can never be determined by one of its

---

95 Bertrand Russell clearly saw this point when he told the following story, ascribed by him to Jourdain: "The situation is analogous to that of Jourdain's chinese Emperor and the nests of boxes. This Emperor attempted to enclose all nests of boxes in one room. At last he

elements, for then it would be self-limiting, and not limited from the outside. A fortiori it cannot, by definition again be its own element. So the concept of *set* may be characterised by the property that a set is something which is not its own element. Therefore the *Russell-class*, containing precisely those sets which are not their own element, adequately characterises the concept of *set* itself. Therefore it cannot be a set, but on the other hand it is formed according to the comprehension axiom, so it *must* be a set. It must be *external to itself*, and that again is an adequate characterisation of the mathematical universe, but again not expressible as a mathematical concept without self-contradiction.

Investigation of the foundations of mathematics by mathematical methods, as it is done in metamathematics or mathematical logic, however, is not completely meaningless. It completes the motion of self-reflection of the second degree of reflection, and thereby clarifies the dependency of reflection of this degree on the principle of structurability. This self-reflective motion - especially in the form of semantics or model-theory - clarifies the nature of mathematical objectivity, the ideal world into which explicit mathematical reflection projects the structures it designs.

Mathematical objectivity can - by definition - be seen *only from the outside*. It is only meaningful if related to something *without* the mathematical realm, either the world of the first degree of reflection, the world of real experience, or a postulating subject.[96] Its definition is externality, for the structurable has to be external to the structuring agent, and the elements of a structure have to be external to each other. The principle of structurability could therefore also be named the principle of externality. And it is characteristic of this principle, that it cannot be understood from the inside. And that is exactly the limit of mathematical knowledge.

2.5 Philosophical reflection

What is left if we abstract from the relativity of the first and second degrees of reflection? Clearly that which is *presupposed* but not explicitly *known* in these degrees. In the first degree we order our experience by means of concepts. This presupposes the possibility of meaningful conceptualisation.

---

thought he had succeeded, but his Prime Minister pointed out that the room constituted another nest of boxes. Though the Emperor cut off the Prime Minister's head, he never smiled again." *An Enquiry into Meaning and Truth*, Chapter 13, part C, Pelican Edition, 1962. p. 189.

96   When prof. M. Mostowski was once asked: "Is everything a set?", he answered solemnly: "I am not a set!".

We can talk about things, plants, animals and people, having certain natures, certain qualities and relations etcetera. *What* we call a thing, a plant, an animal, a quality, a relation, may be exposed to doubt and cultural differences, but *that* they are there, that these ideas may be used meaningfully is an unavoidable certainty of human practise. Such ideas seem to be connected to the ontological nature of the relationship constituting the empirical world. In philosophy they are usually named 'categories,' the most elementary concepts by which our understanding is guided in conceptualizing the world of experience. Aristotle's an Kant's systems of categories, although they differ in ontological status, both pretend to be explicit accounts of the implicit guidelines of conceptualization and judgement. In the following two term: 'perspective' and 'principle' will be used for describing the content and the form of insights enabling us to conceptualize and structure the world of experience. These insights are understood to be already implicitly present in the concepts, judgements and structures we find in ordinary life. Without them the conventions of a culture could never be meaningful and permanent, nor could technology function effectively.

The same holds for the mathematical world of ideal structures. These structures may be created in an infinite number of ways. They may be multiplied indefinitely, and connected in ever more intricate ways. But all this happens on the basis of the simple act of mathematical postulation, which is based on the idea of structurability. And this idea is deeply rooted in the experience that we are able to structure our world. And the possibility of structuring is again connected to the ontological nature of the relationship constituting our practical world.

If we claim something to be relative, we cannot avoid claiming the relationship in question to be absolute.

But is there *one* relationship for each of the two degrees? At least for the first degree this is highly questionable. We do not talk about lifeless things in the same manner as we talk about organisms. We talk about the human world quite differently in comparison with what we say about the world of 'nature.' We have scientific, political, technical, medical 'discourse,' and perhaps even a finer decomposition into indeterminately many 'language-games' has to be made. But also mathematical thinking has many branches, theories, 'spaces,' algebraic systems and number-systems.[97] One abstract relationship, or a few of them, will clearly not do.

In the third degree of reflection, the whole of the experienced world stands before us again. Not in order to be conceptualised or structured, but

---

97 Which even have figured as paradigms for the idea of 'discourse' or 'language-game' in Wittgenstein's *Philosophische Untersuchungen*.

in order to be understood *as* conceptualisable and structurable. In the same way as conceptualizing and structuring, this understanding too is present implicitly in ordinary life. In explicit mathematical thinking, however, it is still implicitly present. The philosophical perspective is now defined as the attempt to make this implicit understanding explicit.

In fact, we now understand that we have been thinking philosophically all the time, be it unsystematically. It also becomes clear that the third degree of *abstraction* was not called so by scholastic philosophy, because it left a certain kind of real *content* out of consideration, such as the individual situation in the first degree, and change and quality in the second. What it leaves out of consideration is only the *form* of conceptuality, and the *form* of structure in the framework of which all content is understood in the other degrees. This means that what 'remains' is precisely the intelligible content which makes these forms function in life, and which gives us the practical certainty of their trustworthiness.

Now if that is so, how can we explicitly say anything at all from this point of view?

We have seen in the other degrees that 'abstracting from,' or 'leaving out of consideration' is not the same as neglecting altogether. In empirical thought we apply empirical concepts in individual situations, and in mathematical thought we shift our attention from one structure to another, judging their properties as if they were observable qualities. Abstraction is rather to be understood as a *polarisation* of our attention. In the first degree between concept and situation; in the second between postulation and structure. And here in the third degree we polarise between absolute and relative experience, between the unspoken and the spoken world.

It is Wittgenstein's mathematical attitude in his Tractatus-philosophy, that makes him *separate* 'Sagen' and 'Zeigen,' the factual and the mystical, what one can talk about, and on what one has to be silent. They cannot be separated however. The unspoken is present in all speech, and we cannot get rid of speech when trying to concentrate on the unspoken, that is exactly the paradox of the last sentence of the Tractatus. But we can speak being *aware* of this relationship, and then we speak differently. We use the same words in different ways, which people who are not accustomed to it find difficult to understand, and rather frustrating. Someone not acquainted with theoretical physics or academic mathematics may be satisfied with the fact that he does not understand a word, because in these disciplines other words or definitions than those of ordinary language are used, which he does not know the meanings of. But a philosopher uses ordinary words in so strange a way! "The principles of being as being"; "Pure being is pure nothingness." Such formulations were regarded as 'meaningless' by many positivists. And in a sense they were right, for in the perspectives of the first and second degree of reflection they do not have any meaning. Hegel, in the preface to the Phenomenology of Spirit,[98] notices that the usual complaint about philosophical language is, that one has to read every sentence twice in order to understand it once. But that is precisely the point of it, because one has to *reflect* to understand such language. It does not serve communication of facts or results, nor postulation of hypotheses or axioms, nor deduction of theorems, but the initiation of a reflective motion of thought. If one is not prepared actually to *perform* this motion while reading, philosophical prose becomes enigmatic or nonsensical. Such reflective use of language has always a somewhat paradoxical character, for its form suggests that a communication is made, but its content is not something directly communicable. It is like pointing. As long as your attention is directed towards the pointing finger, you do not get the point. Even Wittgenstein's "What one cannot speak about, thereof one has to be silent" can be read in this way. For in precisely this sentence he is speaking about it. And yet not speaking *about* it. The sentence expresses precisely the internal tension of philosophical language, which the author at that moment pretended to have resolved for ever.

The third degree of abstraction is traditionally called 'metaphysical.' Metaphysics, however is only the most explicit form of the degree of reflection belonging to it. Reflection of this degree, which means understanding experience in a perspective based on intelligible principles, must be contained already in all thinking, even the most trivial. Philosophy is only the explicit form of this reflection. Just like pure mathematics in the case of the second degree, it is already a self-reflective form of it. Metaphysics is in this respect comparable to mathematical logic: it is the attempt to complete this self-reflective motion. It is therefore directed towards an explicit insight into the realm of principles itself.

So all real philosophy is moving explicitly on the third reflective level. However, is it a *level*? The idea of levels of reflection seems to be rather commonsensical, and one would ascribe such a notion to empirical thought rather than to philosophy. If it were tried to speak of *dimensions*, rather than of levels of reflection, at least the 'geological' connotation would be eliminated, but one would still be looking at the question in a mathematical perspective. Therefore the first problem arising here, is how the degrees of reflection are themselves to be understood philosophically. We cannot properly discuss this problem before we have dealt with the *dynamics* of the matter, which will be done in the next chapter. What can already be said, is that, contrary to common sense and mathematical thinking, philosophical reflection does not produce any external result. No conceptualisation of experience, no ideal world of structures, no 'theory' in the sense of a hypothetical explanation of phenomena. It only produces the internal result of awareness about the

---

98   G.W.F. Hegel, *Phaenomenologie des Geistes*, ed. Hoffmeister, 1952 p.52; See A.V. Miller's Translation: *Hegel's Phenomenology of Spirit*, 1977. It is advisable to read the whole Preface.

relationship of abstraction in all thinking, practice and use of language. This internal result can be expressed in language, in the philosophical use mentioned above. This has produced philosophical tradition and philosophy as a discipline, which is all useless if it does not result in the actual performance of philosophical reflection by the person who studies it. Just like reading literature, in particular poetry, which stimulates fantasy, philosophical reading has to be active reading or else it is worthless. Reading philosophy is itself philosophizing. So as a division of disciplines too, the traditional distinction of the degrees of abstraction belongs to the perspective of the first degree. It distinguishes A) all empirical disciplines; B) all mathematical disciplines, and C) all philosophical disciplines. But then, as in all such divisions, it remains a problem where to draw the borderlines. What with: theoretical physics; mathematical economics; computer science; mathematical logic; methodology of science? This is so, because the *activities* corresponding to the three degrees of reflection as I distinguish them, are not distinct in the same manner as the disciplines. They presuppose each other implicitly within any activity of thought, and are essentially related to each other. A conceptual division of experience is also a result of structuring experience in a certain perspective, based on principles. All mathematical structure has some source in conceptual division and therefore presupposes the principles of common sense besides the principle of structurability. All insight into principles presupposes practical and theoretical involvement with experience. Therefore philosophical reflection presupposes the conceptualizations of common sense and the structural models of science and technology. It also involves the *activities* of conceptualization and structuring, which we see in the necessity of examples from ordinary life and of schematic illustrations, but also in the systematic structure generated by the motion of reflection. On the other hand empirical theory and scientific modelling implicitly express an element of reflection, revealing the meaningfulness of what would otherwise be sheer arbitrary division and construction. The same is true for pure mathematics, which is led by intuition about what kind of structuring is fruitful or essential in relation to the principle of structurability. This can be seen in the recurrence of the same kinds of operations in all the many specialties of mathematics. There is for instance always some generalisation of the operations of addition, multiplication, involution, and their converses, whatever abstract entities, sets, topological spaces, algebraic structures etc. it concerns. Mathematics is not quite so disparate a field as it appears on first sight. A very concentrated form of reflection is at work implicitly in it, giving direction to the investigations. Therefore it is not at all amazing that mathematical inventions such as non-Euclidean spaces and fractal structures find applications some decennia or even centuries after their discovery. Mathematicians direct their attention towards ideal entities, but they do so within the total reality of experience, and aware of its unspoken dimensions. Mathematics is not purely an ideal reconstruction of the common sense- or scientific world, but is based on an original reflection with respect to the totality of experience.

So the three degrees of reflection, considered as degrees of the *act* of reflecting, are necessarily present simultaneously. The disciplines generated by their *explicit* use distinguish themselves by the conscious attempt to polarise attention according to only one of the degrees.

Now what is characteristic of the explicit philosophical form of the third degree of reflection? In the first place it directs our attention towards abstractional relationships, present in all our knowledge. All knowing involves an act of directing attention, and philosophy directs our attention to our doing this, not as a *fact*, but as something transcending all facts in content and form. This is sometimes named the self-reflection of the reflection of the understanding, but as we have seen there are as many forms of self-reflection as there are forms of reflection, so we have to explain this more precisely.[99] Hegel's philosophy can be said to be the most extensive and consistent elaboration of this point of view.

> Considered in their systematic order, his works may be said to begin with the Phenomenology of Spirit, which develops this point of view. This could be called his epistemology. Then we have the *System*, consisting of the Logic, the Philosophy of Nature, and the Philosophy of Spirit, in which he tries to give account of all conceptual, natural, social and spiritual reality from this point of view. As key-stone we have the Philosophy of History, and the History of Philosophy, dealing with the measure in which the philosophical account of reality is indeed recognizable in factual development. Hegel's philosophy has been criticised in many ways. Its developments are said to be arbitrary, or induced by prejudices of his time or of European culture in general. His dialectical method is criticised as being inconsistent or even incomprehensible, the 'absolute Idea' is said to amount to 'looking God into his playing cards,' and his philosophy of nature is found to be contrary to science. His political philosophy is criticised as either too conservative or too liberal. The whole system is found to be 'closed,' and therefore dogmatic, and his philosophical point of view is criticised as 'idealistic.' His theological conclusions are named 'pan-entheistic,' and he is accused to be 'Spinozistic' in his beliefs.
>
> Now there may be something in all this criticism, but it does not take away the merits of his approach, which is the attempt to re-think the reality of his time from the point of view of philosophy. Of course then, this attempt is bound to the politics and science of his time. Of course the system is now experienced as 'closed,' for the age in which is was made is over. Yet, in its philosophical *method* it is no more obsolete than Euclid's geometry or Newton's mechanics are in their mathematical or physical method. Certainly we now have to question - as in the scientific examples - again the *scope* of the method. And in Hegel's case this has much to do with the more theological criticism I mentioned.

---

99   This expression has been coined by J.H.A. Hollak, e.g. in his article Wijsgerige reflecties over de scheppingsidee, in *De Eindige Mens?*, 1975.

Hegel's method is called 'speculative dialectics.' It aims at making explicit the immanent order of the perspectives in which we understand experience. It proceeds by giving a systematic account of the contents of each perspective in relation to its principles, which determine the *form* in which these contents are thought and expressed. This always leads from the naive assumption of immediate adequacy of form and content via the discovery of their mutual inadequacy through an inconsistency, towards insight into their relative opposition,[100] which opens a new perspective, in which a new content is understood in a new form. It is always a relative adequacy which is reached, and partly the motion leads back to an earlier position, now seen in a new light. The traditional structural image for this motion was a spiral, but in fact a fractal spiral, wound infinitely into itself would be more in accordance with what happens.[101] The motion of speculative dialectical thought is very intricate, for there is a continuous change of content as well as of form, and a constant proceeding and returning. Yet we may ask whether it can result in a completed system, as it has apparently done in Hegel's case. Now of course the *oeuvre* of a philosopher can always afterwards be considered as a completed whole, just like a human course of life. Hegel himself used the expression *encyclopedia*, which - in its philosophical intention - means that it moves in a reflective way through all knowledge. But, just as in the case of an ordinary encyclopedia, there is no claim to completeness for now and ever after. The only final claim Hegel is making, is methodological. Now even in this respect one may have doubts. Nowadays one can still develop dialectical considerations about special subjects, such as technology.[102] But nobody would think of writing a new encyclopedia in Hegelian style. Yet Hegel himself considered this the only adequate way: "The true form in which truth

---

100 Relative opposition is the opposition of correlatives as such. See chapter 1, section 4.

101 This was pointed out to me by Dr. Manfred Gies in relation to Mandelbrot's set, which is essentially a circle consisting of circles consisting of circles ad infinitum. Hegel himself describes the motion of philosophical thinking in the introduction to his *History of Philosophy* as 'a circle of circles.' Of course fractal mathematics only gives us mathematical images with which we can *illustrate* the effect of dialectical motion of thought. It has nothing to do with the real philosophical content or aim of this thought. Therefore it is not possible to explain the dialectical motion, or the structure of Hegel's system by mathematical reconstruction. In fact the mathematical image follows from the logic of the dialectics, not the other way round. For the same reason all attempts to formalize dialectical logic are hopeless. It mocks all formula's, because it proceeds by reflection on their possibility. Its structure is only the skeleton it leaves behind.

102 See chapter 1 of this book.

exists, can only be the philosophical system of it".[103] Why is it, that he valued coherence higher than we do?

Now I come to the point where I have to formulate my own hypothesis about this. That is also the point where I leave the safe road of tradition, and try to advance on my own speculative powers. Needless to say that such attempts are fundamentally open to discussion.

It is often thought that the reflections of speculative dialectics move on necessarily and immanently, which means, that they express the development of only one fundamental principle. Hegel himself affirms this by what he writes on the *absolute idea*[104]: "All else is error, unclearness, opinion, striving, arbitrariness and transience; only the absolute idea is *being*, intransient *life, truth knowing itself*, and is *all truth*". Now one may wonder how so rich a material as that which is developed in Hegel's system can evolve from one principle, and in fact it does not. It is taken from experience and the sciences in all their diversity. But not just as empirical material. It is, as Hegel expresses it in an untranslatable way: "Auf den Begriff gebracht", which I understand as: considered in the perspective of philosophical reflection. But what is the difference. Precisely the difference between the first and the third degree of reflection. The same as the difference between what Edmund Husserl calls the 'natural attitude' and 'the phenomenological attitude.' In the natural attitude our attention is directed towards *what* is denoted by our concepts, and this is *reality* for us. In the phenomenological attitude we look at *why* it is so denoted, what contents of experience are hidden in these concepts. We try to look deeper into the experiential foundations of conceptualization. The further question: what is essential in this content of experience leads us to the idea of 'Wesensschau,' the understanding of essential features hidden in the use of concepts. Now this is exactly what Hegel does all the time. It is the fuel that makes his dialectical machine run, and it is the hidden dimension of his system. This too accounts for the fact that in our time, in which the rationalistic search for one fundamental principle has been weakened, or perhaps banished to theoretical physics, the many essences become relatively more important than the tendency towards a comprehensive system. But this raises, on the one hand, the question whether and why we still can 'locally' develop the essentials of a phenomenon systematically instead of being condemned to the disparate descriptions of the 'Bilderbuch' phenomenologists.

---

103   Preface *Phenomenology of Spirit*, see note 23.

104   *Wissenschaft der Logik*, second part, third section, third chapter. Suhrkamp Werke 6, p. 549.

On the other hand we may become curious what is wrong with Hegel's system then, and with his absolute idea.

It is clear, that in order to answer these questions, we have to develop, at least in first draft, a philosophy which gives both the plurality of essences, and the one Idea their proper place. In doing so, some solution to the problem in what sense the one idea may be called 'absolute' has to be found. This also involves transcending mathematism. For the unity of the many essences cannot be structural. A structural unity can *either* be seen as a disparate plurality of individuals, *or* as a single totality. A set, e.g. is either a disparate plurality of elements, or itself an element of some other set. A structure or 'system' in the mathematical sense has either maximal coherence or no coherence at all, and that is just because its coherence is essentially *postulated*.

This is also the dilemma experienced by Wittgenstein in his problem of family resemblances.[105] He felt that a concept had to cover either one single essence or an incoherent plurality of essences. Yet the coherence of the concept had to be explained. So he chose for a partly overlapping, but essentially incoherent plurality, a truly mathematical solution, easily translatable in set-theoretical terms. In fact, it is clear from the way he formulates the *problem*, that it is his philosophical aim to leave this mathematism behind. But he does not succeed for he refuses to enter the realm of real philosophical reflection explicitly. For it is the realm of silence, which he chose not to break.

We have seen in Chapter 1 that there are two other senses besides the structural one, in which a whole can be understood as made up by its parts. It can be divisible into parts which are not constituent of the whole in their own right, but only in so far as they are actually contained in the whole. Examples were a chemical compound and an organism. Outside the compound, the elements have quite different properties than they have as actual parts of it, and outside the organism, the parts usually cannot even live. The other sense was that of a whole of which the parts cannot be individuals in themselves. The most clear example of this is a relationship. It does not consist of anything else but its correlatives, but in so far as they are correlatives, one cannot even think of them as separate individuals. It is true of course that a parent and a child are separate individuals, but not the parent's parenthood and the child's childhood. So with part and whole, subject and object, unity and plurality, identity and distinction etc. In a sense the correlatives as such are parts of the relationship, but it is sheer nonsense to even think of separating them from it.

---

105   See H.P. Boukema, Familiegelijkenissen. Wittgenstein als criticus en erfgenaam van Frege, 1987, pp. 42-70.

We may conclude from this, that it is not necessary to think of a whole from the structural point of view only. We can e.g. think of Hegel's system as organic. The one Idea giving life to it, but not explaining all its details, just as the life of an organism does not explain its precise organic composition. We can also think of the system as purely relational, all its elements being determined completely by their mutual relationships, completely inseparable from the whole. Of course this does not solve our problem, for there are so many good arguments for both views, that we do not know which one to choose. We shall have to dive deeper into the problem of unity and plurality, and take the risk to come up muddier. But that will be postponed to Chapter 5.

Concluding this chapter I have to make a remark about the relation of philosophical reflection to ordinary human life. Conceptual and mathematical reflection seem both connected to activities which pervades the whole of human existence. Our life-world is a conceptualized world and all our technical practices have to do with structuring. It is not often recognized that this is also true for philosophical reflection, although everybody will recognize it *for him or herself*. We all have the feeling that there is an unconceptualizable and unstructuralizable element in our life, and we experience that we are *thinking* about it. We make use of some terminology, either borrowed from a discipline we have learned or from quotidian language, but we are constantly aware that we use it in a Pickwickian sense. Most people do not discuss such 'elusive' thoughts with others, because they feel that the language they use is rather 'private,' although if they did discuss it, some people would in fact recognize what they were talking about. Those who have developed a personal way of speaking about such thoughts and the experiences they originate in, sometimes are called 'wise.' They have contact with a deeper dimension of life and seem to have intuition about how to act best in certain difficult situations. We may use the term 'existential' for this sphere of life, which everybody somehow experiences, and which in a sense transcends the sphere of common sense[106] and technical practice but is not philosophical in the disciplinary sense.

This dimension of 'wisdom' in ordinary life is the germ of explicit philosophical reflection. We all know and understand more about life than is made explicit in our common-sense notions and scientific theories, although it is presupposed in them. This surplus of knowledge does not necessarily lead to philosophy. It finds expression in practical action, artistic creation, religious

---

106 Erwin Edman, in his book *The Philosopher's Quest*, introduces the term "Uncommon sense."

experience, and in all experiences of value in whatever sense. In these domains it is difficult to distinguish the good from the bad, the pure from the sham, the clear from the foggy. In philosophy this is sometimes equally difficult, but its *aim* at least is clarity.

One could ask here, what is the use of making explicit what is already implicitly functioning. This question will be dealt with more extensively in chapter 5. What can here already been suggested is that the same question can be asked concerning mathematics. If we are already always structuring the world, what is the use of explicit and ideal structuring. In fact this use often goes unrecognized by those involved in practical work in science and technology. The results of mathematics are taken for granted and mathematical research is regarded as esoteric business. Yet it is clear to those who know the least about the history of science that the impact of mathematics on the development of human knowledge has been enormous. In my opinion the same holds for philosophy.

The important conclusion here, however, is that philosophy is not an isolated discipline, but the enterprise of bringing to light a rich layer of knowledge which is already contained in ordinary life, in what I have called the world of experience. This is the reason why people are interested in it at all. They recognize something of their own experience in it. Yet this experience itself always remains infinitely much richer than what can be made explicit in philosophy, and this is a deeper meaning of the dictum: *primum vivere, deinde philosophari*

Chapter 3
The degrees of reflection: examples of the dynamics

In Chapter 2 the three degrees of abstraction are distinguished by the levels of reflection they primarily produce. These levels are pictured there as being completely independent of each other. In fact, in all knowledge or thought all three are involved in a certain relationship, which may be quite complicated, for reflection is a motion of thought which gives rise to many different perspectives.

The relationship between the levels of reflection may be unstable, so that a shift of relative weight takes place. The mathematistic tendency of modern science can be described as a continuous shift of weight towards the second (mathematical) level. On the other hand the various forms of criticism of mathematism tend to shift weight just in the opposite direction, the effect not being equilibrium, but ambiguity. In this chapter various relationships and shifts will be exemplified by some recognizable ways of thinking, in philosophy as well as in science.

3.1 Between empirical and mathematical reflection

As we have seen in Chapter 2, all empirical knowledge presupposes a division of the world of experience according to the categories of the language we use. From the point of view of mathematical thinking, this division must be regarded as a certain *structuring* of this world. This structuring presupposes meaningful perspectives which we cannot do without, for if we try to deny such meaning we fall into the antinomies of scepticism. But the idea of meaningful perspectives points to the third level of reflection. Therefore real empirical investigation involves all three levels of reflection. If we try to be *empiricists* and ascribe prevalence to the first level, we cannot do so without implicitly recognizing the other levels. And if we want to lay a foundation for our empiricism, the other levels are bound to rise above the surface of consciousness.

Some examples will be given of the motion of thought from the empirical to the structural level of reflection, sometimes even touching the metaphysical level, but always bound to return to its origin and repeat the process.

The rise of modern science

This motion is present at the beginning of modern science. Galileo, in his *Dialogues on the Two Main World Systems*, continually appeals to common sense and experience, when in fact he is reasoning mathematically.[107] Until the nineteenth century the distinction between empirical and mathematical arguments was not clearly recognized. A continuous shift of weight between the two levels took place, but the ultimate victory was on the mathematical side, as the longterm development of science proves. This shift was directed from qualitative distinction towards measurement, from conceptualized experience to structuralized experience. Experience became progressively analyzable into 'effects,' due to certain laws, which by their combination in the specific case determine what happens. E.g. the effects of gravity and friction in mechanics, the effects of different forces or fields or even effects computed in different theories, e.g. classical mechanics and electrodynamics, or nowadays quantum mechanics and the theory of relativity. Effects were first ascribed to forces, then more and more to laws. There was also a continuous striving for unification, because it was implicitly understood that all analysis is structuring, so it must be possible to conceive of *one* structure containing all models of effects as substructures.

Structuralism

Even the idea, so popular in our time, that it is the structure of language or of culture that determines how we experience the world, is an example of the shift of weight from the empirical towards the mathematical. There is a persistent ambiguity in this way of thinking. I shall illustrate this by the case of structuralistic interpretations of de Saussure's thesis of the arbitrariness of the relation of *signifié* and *signifiant*, of 'meaning' and 'sign.' It can be understood to mean that there need not be a relationship as regards contents between what is used as a sign, and what is understood to be its meaning. This is an obvious fact, already noticed by many investigators. Hegel expresses it by saying that an alien soul has been embodied in the sign like a Pharaoh in a pyramid.[108] This thought *presupposes* that what is used as a sign and what

---

107 E.g. in his argument about the ball on an inclined plane (Second Day) in which he deduces a principle of inertia. Sagredo makes Simplicio believe, that the argument is purely common-sensical, whereas in fact it is a mathematical continuity-argument in an idealised case. *Dialogo sopra i due massimi sitemi del mondo*, Edizione nazionale VII, English translation see Lit. list.

108 Das *Zeichen* ist irgendeine unmittelbare Anschuung, die einen ganz anderen Inhalt vorstellt, als den sie für sich hat; - die *Pyramide*, in welche eine fremde Seele versetzt und aufbewahrt ist." [The *sign* is some immediate impression, which denotes a content

is meant as its meaning already have a content of their own. In a structuralistic interpretation, this content becomes nothing but a position in the structure of language or the structure of experience, these two being essentially the same. The point of view - here called 'empirical' - in which the world of experience *has* certain structures, dependent on the real experienced properties of its elements, undergoes an about-face, and turns into the mathematical point of view, in which the structure comes *first*, and qualitative properties are the effects of structures. Although structures are still thought of as *real*, in fact they become *arbitrary* in the sense that there is no reason for preferring a particular structure. Such a reason could be found in the properties of its elements. But in the structuralistic order such properties are now understood as *effects* of structure itself, which makes this solution circular. If one explains, for instance, the behaviour of people from their place in the social structure, one cannot at the same time explain the specificity of a social structure from the behaviour of people. Therefore in such a perspective any structure - if sufficiently sophisticated - would do, and the meaning of this *specific* way of structuring the world disappears. It cannot be understood in another way than as originating in the power of the prevailing culture-bearers, ultimately based on the threat of violence. To speak otherwise than the prevalent discourse prescribes, is to go against ruling powers, and the question whether this is genius or nonsense loses its urgency. Here again is a necessary shift from mathematical reflection back to empirical reflection: mathematics has no reason to offer for the prevalence of one structure over another, so this prevalence is now reduced again to a brute fact of power.

The methodology of empirical science

In natural science the overall structure of the universe cannot be deduced from the principles of natural science itself, but has to be recognized as a fact. All laws have boundary conditions, which have to be specified in order to apply the laws. So natural laws structure the world of experience only relatively, with respect to certain factually measured or prepared quantities. It might turn out that with the given laws there is only one possible overall structure of the universe, but even then the laws themselves, the nature of the parameters figuring in them, and the values of the natural constants cannot be justified mathematically. The determination of the natural laws governing (relative) structuring can be understood in three ways. It can be conceived of as empirical, as hypothetical or as necessary.

---

completely different from its own; - the pyramid, in which an alien soul is introduced and kept.] Enzyklopädie § 458 (Anm.), Suhrkamp Werke 10, p. 270

The first conception leads to the view that the role of mathematics in natural science is primarily methodological. Science is understood as fundamentally empirical. The question why this method has been so successful, however, remains unanswered.

The second conception is similar to structuralism. The experiential reasons why the hypotheses of science were *meaningful* have disappeared, only the *fact* of a certain structuring remains, and as this fact cannot be deduced theoretically it has to be justified empirically. But because the original empirical content has been repudiated as belonging to the 'context of discovery' or pre-scientific knowledge, the empirical basis has to be looked for elsewhere, e.g. in sociological considerations about the process of scientific investigation.[109]

The third conception, now rather popular among theoretical physicists, leads us away from the contingency of the first two levels of reflection, towards the third, which deals with the absolute. But a little bit too fast. We have a kind of metaphysical shortcut here, in which all contingency is reasoned away in one stroke. What happens in the world is determined by an absolute structure, installed by some deistic God, about whom nothing else can be said than that He established the laws and natural constants as they are.[110]

Another striking example of this kind of metaphysical shortcut is the anthropological doctrine of René Girard, which - in its most extreme form, defended by Jean-Michel Oughourlian - tries to explain all social and psychological phenomena (and in some cases even *natural* phenomena) by the law of *mimesis*: 'What a person desires is determined by what is desired by another person in the measure of the social importance of the latter, and in a measure strongly increasing with diminishing social distance between the persons.' Oughourlian compares this to Newton's law of gravitation: all masses are attracted by other masses in the measure of their quantity and of the inverse square of their distance. Now it is a fact - and the works of Girard himself and of his followers[111] testify to it - that this idea sheds a new and revealing light on many phenomena in the human world. This success of the doctrine is however hampered by the

---

109 In this perspective structuralists such as Foucault and Latour have gratefully taken up the thread of Kuhnian sociology of science, although Kuhn himself denies that this was his intention.

110 This can be compared to the position of Wittgenstein in his *Tractatus*. Only there the absolute structure is not physical but *logical*. In the *Philosophical Investigations* he seems to return to the first conception, akin to structuralism, in which the structures of language use are determined by the grammar of language-games embedded in contingent forms of life. Just as in the case of structuralism the question of what makes these games *meaningful* cannot arise in this context.

111 See R. Girard, *La violence et le sacré* and the other works mentioned in the literature list.

desire to make it *universal*. It is contrasted with the - according to Girard *romantic*[112] - idea that our actions come from ourselves. As if in mechanics, mass were not regarded as a *property* of matter itself, determining its interaction with other masses, but only as a quantity determined by the relations between elements of a mechanical *system*. This is of course a legitimate and sometimes advantageous point of view, just like the structuralistic thought that meanings are determined by the totality of linguistic relationships. It approaches the mathematical perspective, explaining everything from structure. But we are in the sphere of real experience here, so one may legitimately ask where the structure comes from. What gives the networks of mimetic connections the structures they have? In order to answer this question, there is nothing to do other than to return to our common sense ideas about people's behaviour, unless one wants to be dogmatic either as an empiricist or as a metaphysician. In the first case the facts need no explanation, and in the second the totality of mimetic relationships points towards a self-structuring process of the universe, acting either as an unmoved mover or as absolute, infinite substance.[113]

We ascertain an oscillating movement of the kind of thinking primarily interested in the concrete world of experience. It moves towards the mathematical level, but it has to return to earth, because it is there that its primary interest lies. Two poles are trying to distract it from the world of facts: structure and meaning. They seem to be alien to this world, and yet play such an important role in it that, without them, there would be no world of facts. What happens when we remove our thinking to a world of pure structure, and look from there to facts and meaning?

3.2 Mathematical views of experience and philosophy

The transition from the first to the second level of reflection is not dialectical, in the sense that consistent development of the first level forces thought by self-contradiction to shift its perspective towards the second level. The first level has a right of its own, and thought can remain there as long as it recognizes the legitimacy of the other levels. And probably it will implicitly or explicitly do so. In the implicit case, it is unconsciously attracted towards higher levels, as we have seen. Empirical thinking will either return from its strayings into the mathematical or the metaphysical, and regard some factual datum as an ultimate reason, or become conscious of its non-empirical dimensions, and turn into mathematical or metaphysical thought.

This transition has the character of Edmund Husserl's εποχη, by which we perform the transition from the natural towards the phenomenological

---

112   Cf. his first important book: *Mensonge romantique et vérité romanesque*, 1961.

113   Look however at P. Tijmes, The genius of the master lies in his limitation, 1994.

attitude.[114] There, we place the reality of things between brackets. We do not give up our actual belief in this reality, but we make an agreement with ourselves to leave it out of consideration, and only look at the determinations of phenomena as such.

In mathematical thinking we place everything which is necessarily connected with sense impressions, e.g. quality and change, between brackets, and only look at structure as such. Even the *act* of structuring and the *principle* of structurability themselves, are left out of consideration. Only their *result*, structure, is taken into account. Therefore we enter an ideal world of structures, in which reasoning and understanding, led by mathematical intuition, constitute the methodological perspective. In this attitude, everything is judged and understood by its possible structures. What Henry David Thoreau seems to have said about truth, also holds for mathematical understanding: *"She looks broadcast over the field and sees no opponent."*[115] Nothing can exist other than what is completely determined by structure. Yet, as in the empirical attitude, straying into the other levels is possible. But having its centre of gravity at this level, the mind always returns to structure as its ultimate reason. This is the kind of attitude which I call *mathematism* in the case that there is no consciousness of its one-sided character.

It is said that for the pure, everything is pure. Nietzsche paraphrases that for the swine everything is swine. But indeed for the structuring mind everything is structure. This idea of structure is so strong and clear, that no inkling of a doubt needs to appear even in the face of the most flagrant counterexamples. Space is structure, so is time. Matter and physical process are structures. So is life, and even the human spirit can be understood as the structure embodied in a computer-program. Nowadays some people even believe that God is structure, the Master Program of the universe. You cannot prove the contrary - try it, and you'll see that you are trying to find the structure of non-structure!

This attitude has been extremely fruitful for science. Since the rise of science it has been connected intimately with mechanistic views, but in the beginning of the twentieth century it equally fitted Ostwald's energeticism.[116] It is not only characterized by the search for measurable quantities, but also by their representation within a purely mathematical framework. Considering

---

114   See Edmund Husserl, *Logische Untersuchungen II*, 1950.

115   From Theodore Dreiser, *The Living Thoughts of Thoreau*, 1958, p.58.

116   Cf. Casper Hakfoort's paper: 'Science defied: Wilhelm Ostwald's energeticist world view and the history of scientism,' 1992.

its success it is not at all amazing that the mathematical attitude has become the principal paradigm of thought, and that it has pervaded strongly the other levels of reflection.

I shall give some further examples of this pervasive tendency. First an example of an attempt at genuine pan-mathematism, then one of the reduction of fact to structure, and finally of mathematism in philosophy.

Mathematical reason has the tendency to try to explain everything, the empirical as well as the metaphysical. This is clearly to be seen in the philosophy of L.E.J. Brouwer, which starts with consciousness, understood as a self-structuring process, and ends up with considering the whole world of experience as a complicated system of causal chains which are purely structural orderings of facts.[117] This is really a mathematician's philosophy, with all the clarity and elegance belonging to it. Yet it is a curious one, for Brouwer does not really believe in the existence of structure in the mathematical sense as I have described it in Chapter 1. There it was argued that mathematical thinking abstracts from change and potency. What is thought of in a mathematical way is thought of as purely actual. Brouwer, on the contrary, understands the mathematical universe as a *process* of mental construction. A process, which is never completed, but contains reflections on its own incompleteness. The reason for this is that he does not see the activities of the mind as *intentional*, as motions of attention with respect to a reality which is ultimately not *caused* by them. He considers the process of the mind as the actual *production* of its constructions. The constructions are understood as mental realities, not as ideal structures. So these constructions are always finite and always growing, essentially by reflection on the process of growth itself. Curiously enough however, this reflecting mind does not see or express its own doings in any other form than its structural results. This means that in fact all reality is reduced to structure, which is in the perennial state of being under construction. Therefore, Brouwer's philosophy does not reach pure mathematism after all. It is rather the hypostasis of the self-consciousness of the mathematical mind, which cannot succeed in denying its own activity.

A probably unexpected but nevertheless striking example of the 'downward' motion, in which the *empirical* level is interpreted mathematically, is logical positivism. Inspired by mathematical logic, it tries to embed all science and philosophy in a mathematical context. Facts are reduced to logical atoms, e.g. protocol sentences, theories to formal systems, philosophy is

---

117 Brouwer's views are discussed more extensively in Chapter 1, section 3.

reduced to metamathematics. Time and space are seen as mathematical dimensions, physical properties as values of parameters, meaning as the existence of a verification-*method* (not of actual verification!). Logical positivism is in a certain sense a reaction against neo-Kantianism, with which it shares the prevalence of a mathematical perspective. It rejects, however, the idea of a synthetic a priori, i.e. of transcendental principles which are fundamental for the possibility, not only of experience, but also of mathematical thought. The development of mathematical logic raised the hope that this discipline could explain everything which before had been considered as transcendental. So the whole transcendental sphere could be reduced to mathematics too, the only a priori being analytic, i.e. logical, and, via the mathematisation of logic mathematical.

This is worked out even more consistently in critical rationalism, in which the concept of verification is stripped from its last empiricist element: the so-called bare facts underlying the protocol sentences. According to this view, we can only test hypotheses which predict certain structures to be found in the phenomena. And if the phenomena *resist* being structured according to a certain hypothesis, this hypothesis tends to become falsified. An experiment is an attempt towards such structuring, and it is crucial if failure of the structuring has important consequences for the whole fabric of a theory. This implies that a falsifiable theory excludes certain possibilities of structuring phenomena. Therefore falsifiability is a purely mathematical property of a theory. A tautological theory, for instance, is not falsifiable because it does not prescribe or forbid any structure.[118]

In both streams of thought *demarcation* plays an important role. Untestable principles have to be excluded from the domain of science, for they are *dogmatic* in nature, and tend to restrict the space in which science is allowed to develop. Such principles are sometimes named 'metaphysical,' sometimes considered to be meaningful for other domains than science, but always excluded from science. For the mathematical paradigm of science, this position is completely justified. We have seen that the introduction of an *explicit* formulation of the principle of mathematical thinking into the sphere of mathematics, Cantor's axiom of comprehension, leads to inconsistency. Mathematical objectivity is by definition a domain *outside* its own principle. But it is also clear that the explicit *statement* of such demarcation criteria precisely for this reason must fall under its own verdict, and therefore is as

---

118 In his *Proofs and Refutations* 1976, Imre Lakatos clearly depicts this paradigm within the mathematical sphere. That it is possible to do so does not prove that mathematics is in any sense an empirical science, but only that empirical science is understood now from the paradigm of mathematics only. Cf. also C. Dilworth, *Scientific Progress*, 1981.

unscientific as Russell's vicious circle principle is unmathematical.[119] So in fact positivism and critical rationalism are themselves forms of metaphysics in their own sense of the word. They do not admit, however, a metaphysical *investigation* of their principles, but fall back into the mathematical attitude by *postulating* their principles, or even, as in the case of those who believe themselves to be followers of Kuhn's philosophy of science fall back to the empirical-sociological level as soon as one asks for the nature of these principles. In the meantime, history of science has revealed that all great scientists had a priori ideas concerning the nature of the phenomena they studied, ideas which philosophers of science have set themselves the task of investigating.[120]

An example of the 'upward' motion is metamathematics, in which philosophical reflection is seen in a purely mathematical perspective. Mathematical thinking reflects on its own doings, but not, as in Brouwer's case *as doings*, but purely with respect to their structural *results*. This gives a self-consciousness to the mathematical mind which is quite different from Brouwer's. Whereas Brouwer sees a continuous process of construction, metamathematics only considers structural properties of theories and interpretations postulated as being already completely constructed. This characteristic of metamathematics is often neglected in textbooks. In order to make the subject accessible to students, syntactic structures are described as structures consisting of *symbols*. In the beginning these symbols are regarded as shorthand notations for well known concepts such as the logical connectives, quantifiers, mathematical predicates and relationships, equality and inequality. Later it is said that they can also be regarded as 'meaningless signs' united in a combinatory way in order to form a syntactical structure. Never mind that the idea of a 'meaningless sign' is self-contradictory -a sign cannot be meaningfully defined other than as: '*Something having a meaning.*' In fact syntactical structures are ordinary mathematical structures. The idea of constructing them is induced by the notational conventions in mathematics,

---

119 This has been elaborated extensively in the writings of P. Vàrdy. Cf. Some Remarks on the Relationship between Russell's Vicious-Circle Principle and Russell's Paradox,1979 pp. 3-22 (Also see my article following Vàrdy's, and Paul Bernays' comment on both articles). Cf. also P. Vàrdy, Zur dialektik der Metamathematik, 1987.

120 As W. Whewell and N.R. Campbell already did in their time. This tradition of philosophy of science has been continued by philosophers such as Harré and Dilworth. Recently, however, these attempts towards *systematic* reflection on the reality of science seem to have been swept away by a flood of *historical* investigations which, however, generally do not contradict their systematic results.

and they are used for *reconstructing* certain aspects of mathematical reasoning.[121] They are 'generic systems,' which means that they are thought of as consisting of all elements obtained by the iterative application of finitary rules, starting from a set of basic elements. The finitary character of the rules, the iteration-process and the basic set are remnants of the idea of the actual manipulation of signs, but it is easy to abstract from the actual use of these signs and concentrate on the structure. The natural numbers form the most simple generic system, having one basic element (0 or 1, as the case may be) and one unary operation (the 'successor-function'). The axiom of complete induction corresponds to the 'closure-clause' in the rules for generic systems: 'The system only contains elements linked to the basis by a finite (or transfinite of a certain order, as the case may be) chain of applications of the operations.' This means that the system is thought of as *completed*, i.e. as a mathematical structure.

In *semantics* the original mathematical meaning of the notation taken as a paradigm, or a variant of it, can be restored artificially, by explicitly mapping the relevant elements of the system into a universe of mathematical objects. This gives a reconstruction of the two aspects of mathematical notation: sign-manipulation and mathematical meaning. By this artifice, the mathematical properties of the relationship between these aspects can be investigated by varying diverse elements of the reconstruction. Mathematical reflection hereby inspects its own doings by its own means.

Although Frege may certainly be named the 'father of mathematical logic,' the metamathematical method as described above is due to Hilbert. He designed it in order to prove the consistency and completeness of elementary arithmetic, so that this part of mathematics could serve as the 'rock bottom' foundation for all the rest. The well known incompleteness theorems, first proved by Kurt Gödel, state that the result of Hilbert's attempt is firmly negative. There is no consistent and complete axiomatizable theory, strong enough to serve as a rock bottom foundation of mathematics. Alfred Tarski added a theorem amounting to the statement that semantics is not reducible to finitary syntax, and Alonzo Church and Haskell B. Curry proved that logic is not reducible to combinatory operations.[122]

The method of all these proofs is essentially the same as the reasoning in the liar's paradox, Russell's antinomy or Cantor's proof of the uncountability of the continuum. The latter proof can be given by considering the diagonal of an infinite matrix; for which reason the method is called *diagonalisation*. I shall now deal with this method somewhat more extensively as it constitutes

---

121   For a rigorous criticism of the tendency to substitute reconstructions for the real phenomena, see Sören Stenlund, *Language and Philosophical Problems*, 1990, Chapter 1.

122   See Haskell B. Curry, *Combinatory Logic*, 1968, Chapter 8. The historical remarks in this book and in Curry's *The Foundations of Mathematical Logic*, 1963, are very revealing with respect to the development of the subject.

a kind of mathematical image of philosophical self-reflection. Therefore it enables us to compare mathematical and philosophical reflection very accurately, which we shall need later in order to overcome mathematism in metaphysics, precisely by taking advantage of the results of metamathematics.

Cantor's original proof of the uncountability of the continuum is purely geometrical. He constructs an infinite sequence of intervals, starting from the unit-interval [0,1] (which we call I_0). If we have an infinite sequence of points P_0, P_1,...P_n,... each next interval I_{n+1} is defined as being contained in I_n, being in length less than half of this preceding interval, and not containing P_n. The intersection of all intervals now defines a point which cannot coincide with any P_n. As this holds for any countable sequence of points, there can be no such sequence containing all points of [0,1], which therefore, considered as a set, must be uncountable. Whether one draws this conclusion, or the other possible conclusion - actually drawn by Brouwer and Weyl - that there is something wrong with considering a continuum as a (completed) set, is not important for the idea of this method.

The 'diagonal' form of the method is obtained by using a fixed partition system of the interval, e.g. the system suggested by the decimal number system. This provides a partition of the interval, first into 10 parts, then into 100, etc. If we take for I_{n+1} an interval of the n+1-th decimal partition, disjunct from P_n, this amounts to changing each n-th decimal of P_n in such a way that the point determined by the sequence of these changed decimals differs from all points of the sequence P_1, P_2,....etc. This point is essentially the same as the point contained in the intersection of the intervals I_0, I_1,...etc. The sequence of the n-th decimals of the points P_n can be regarded as the diagonal of the infinite matrix of which the n-th row consists of the decimal-sequence of P_n.

But this 'diagonal' form of the argument can be transformed one step further into a 'functional' form. For this purpose we must notice the fact that an infinite sequence is essentially a *function*, mapping the sequence of natural numbers into a certain domain, e.g. points of [0,1], natural numbers or other mathematical objects. So the symbol 'P' we used above in expressions of the form P_n, can be regarded as denoting a function which maps the natural numbers into [0,1]. But a point of [0,1] is determined by an infinite sequence of decimals, which is essentially a function mapping the natural numbers into the set {0,1,2,3,4,5,6,7,8,9}. If we denote this set of 'decimals' by **D**, and the set of all natural numbers by **N**, we can use the notation P(n) for the function mapping **N** into **D** in such a way that P(n) is again a function and P(n)(m) is the m-th decimal of the decimal sequence of P_n. We write P(n):**N** → **D**. Now P itself stands for a function mapping **N** into the set [**N** → **D**] of all mappings of **N** into **D**. The set [**N** → **D**] actually corresponds to [0,1] as a point-set. So we can write P:**N** → [**N** → **D**]. We could also regard P as a function of two arguments on **N**, but for our purpose it is more convenient to think of P as a function on **N** with as values the decimal sequences of the points P_n. Now let C be a function defined on **D** with values also in **D**, such that for d=0,1,...,7 C(d)=d+2; for d=8 C(d)=0, and for d=9 C(d)=1, so that C changes the decimals cyclically. Now our diagonal argument can be reformulated as follows. C(P(n)(n)), regarded as a function of n defines a point in [0,1] which is different from any point P_n, because |C(P(n)(n))-P(n)(n)| > 1 for all n.[123] In other words, if the sequence of decimals for which the n-th decimal is given by C(P(n)(n)) were to belong to any point P(m) there would be a contradiction, for then there would be a decimal sequence for P_m, for which C(P(m)(m))=P(m)(m) would hold, which is impossible because C is constructed in such a way that for no d C(d)=d. So C cannot have a *fixed point*. A fixed point of a function

---

123 We have chosen C in such a way that the effect of the ambiguity in decimal notation for rational numbers has no effect on the result.

is a value which is mapped onto itself by this function. So p is a fixed point for F, if and only if $F(p)=p$.

The diagonal argument is now seen to amount to the indirect proof which derives a contradiction from an assumption by the construction of a fixed point of a certain function defined in such a way as to have none. In fact Russell's antinomy may be similarly reconstructed as the proof that the axiom of unrestricted set-comprehension leads to a fixed point for the negation operator, which of course by definition cannot exist either. In the same way, the proof of Gödel's theorem involves the construction of a formula which is true if and only if it is not derivable. In a certain, somewhat more complicated, sense this formula is a fixed point of the predicate of non-derivability. The construction of such fixed points by the 'diagonal argument' always proceeds according to a simple scheme, which will be explained now by another consideration of the transformation of Cantor's proof given above.

We now give the proof in its indirect form, starting from the hypothesis that each point of [0,1] is a P_n for some natural number n. That means that also the point Q, determined by the decimal sequence $C(P(m)(m))$, is also a P_n for some n. We mane this special natural number g, so that for every natural number m it is true that $Q(m)=C(P(g)(m))$, which means that the m-th decimal of the decimal sequence of Q is equal to the m-th decimal of the decimal sequence of P_g. As a consequence of this, it is also true that $Q(g)=P(g)(g)$, which means that the g-th decimal belonging to Q is the same as the g-th decimal belonging to P_g. But by the definition of Q we have $Q(g)=C(P(g)(g))$, so it follows that $C(P(g)(g))=P(g)(g)$, which is impossible as a consequence of the definition of C.

We notice, that in this proof there is a one-to-one correspondence between the natural numbers, the points P_n and the functions P(n). So we may simplify the reasoning by regarding the natural number n itself as a *function*, mapping **N** into **D**, defined by the decimal sequence of P_n. This simply amounts to identifying P(n) with n, and it leads to a purely notational simplification of the proof. By definition $Q(m)=C(m(m))$, and therefore $Q(g)=C(g(g))$, but $Q(g)=g(g)$ for Q=g (as a function, not as a point), so $g(g)=C(g(g))$. If we now forget all about points and natural numbers, and assume that all signs we use denote functions, we realise that we have proved that *every* function has a *fixed point*. Of course we have done something here, which is usually not allowed in mathematics. We have applied functions to *themselves* as arguments. But if there is a one to one correspondence between a set of functions and the domain of their possible arguments, we in fact may apply the functions in a Pickwickian sense to themselves. Leaving out the Pickwickian character of this self-application is only a question of convention.

Now we have found the core of the diagonal argument. This reduction to its core is essentially due to Church and Curry, and there is a lot more to say about its structural consequences. But here we are not interested in technicalities, but in the philosophical implications.

The core of the diagonal argument amounts to the following simple 'computation':

$g(m)=C(m(m))$ by definition for all m

so $g(g)=C(g(g))$ and $g(g)$ is therefore a fixed point of C.

We see that g is defined by self-application, and is subsequently applied to itself. The trick is *self-application of a result of self-application*. The argument can be simplified one step more, but it thereby becomes mathematically useless:

$g(m)=m(m)$ so $g(g)=g(g)$.

Not a very surprising result, yet very interesting, for what happens here? We substitute g for m, as a special value of this general parameter. If we 'mark' the *substituted value g* by printing it in italics, the result becomes $g(g)=g(g)$, and we see that *g* as an argument is transformed into *g* as a function, taking the place of g.

A function is an ambiguous kind of mathematical object, as Wittgenstein already remarked.[124] It may be thought of as an ordinary mathematical entity, but sometimes it is dealt with as if it represented an *action* or operation, e.g. a transformation or a mapping. Now the category of action is alien to mathematical objectivity, and everything is done in order to remove it from this sphere. A function is e.g. regarded in set theory as a set of ordered pairs. But what determines this set? A set-forming *operation*. Can we also eliminate it? Yes, we can reduce it to the application of an axiom, stating the (conditional) *existence* of a certain set. But how about *applying* the axiom? Yes, that we do, but that is purely subjective. The set is thought of as existing independently of our noticing it to exist in our reasoning. So in fact the action of a function is really the action of the *subject of mathematical reasoning* projected within the sphere of mathematical objectivity. Mathematical abstraction starts from our action of *structuring* by separating in thought the result from the activity. The results constitute the timeless sphere of mathematical objectivity, the activity on the other hand remains on the side of the subject, which locates itself *outside* this sphere. Yet everything within this sphere is its product; every structure is the result of structuring and has acquired its unity and individuality by postulation. Every mathematical object is created after the image of the subject, and on the other hand it is completely the opposite of the subject, completely external to it. This ambiguity is part of the nature of the subject of mathematical reasoning, so on the one hand the objects can be regarded as images of the subject's *activity*: unities add themselves to other unities, numbers add to other numbers or multiply them or raise them to a power, functions 'act' on arguments etc., while on the other hand it is the *subject* who does all this, and the objects are willy-nilly manipulated.

In modern mathematics[125] this activity is again objectified as a mathematical operation and later as a function, which still has to be *applied* though. However much of the subject's actions is projected into the sphere of mathematical objectivity, the *real* activity always remains with the subject itself. The projecting light as such cannot be projected. But that is precisely what is shown in the diagonal argument. It produces a fixed point of a function, which within a certain domain or under certain presuppositions cannot have a fixed point. Therefore it shows that something exists *outside* that domain or something violates the presuppositions. For Cantor's proof it produces a point outside the given sequence, violating the presupposition of countability. For Gödel's proof it shows a non-decidable formula. For Russell's antinomy it shows something outside the universe of sets. In fact, the fixed point is a new projection of the subject, which can always be found, for the subject itself is essentially outside *any* mathematical domain. Now we can also understand why it is by self-application of a result of self-application that the construction of the fixed point takes place. Application is the projected image of the activity of the subject in the sphere of mathematical objectivity. So self-application is the image of its *projecting* activity itself. If a

---

124 Ludwig Wittgenstein, *Remarks on the Foundations of Mathematics/Bemerkungen über die Grundlagen der Mathematik*, 1964, Part IV, pp. 133-157; or *Bemerkungen über die Grundlagen der Mathematik*, 1974, Teil V, pp. 257-302.

125 Archimedes could not introduce the concept of momentum in his theory of the lever, because quantities of different kinds (length and force) could not be multiplied. Multiplication then could not yet be objectified mathematically as an operation or even a function, as gradually became possible in modern times. An infinite sequence could not be objectified for the same reason in Archimedes time. Therefore he could not develop a theory of limits, necessary for the infinitesimal calculus, although he had developed its technical methods. A limit depends on an infinite sequence *as a whole*, and such a sequence is essentially a function, which in Archimedes time was still felt to be something *we do*, not as something *objective*.

function (the function C in our example) is applied to this image, we get another image of the same activity. The image is transposed, but the subject itself is of course indifferent to this external transposal. The self-application of the function expressing the transposal (g in our example) represents the subject's indifference as a fixed point of the function (C) which causes the transposal. If this function is the 'identity function' (for which the value is by definition equal to the argument), the result g(g)=g(g) teaches us that mathematical equality is itself the projected image of the subject's indifference. This is affirmed by mathematical practice, in which equality is always obtained by identification, which means that an equivalence class of objects is, by postulation, regarded in a certain context as being only *one* object. Thereby we presuppose that the nature of the context is defined by the indifference towards the distinctions within the equivalence classes. If we calculate 'modulo 3' the distinction between 2 and 5 is indifferent to us for this way of calculating. And this indifference precisely *defines* this way of calculating.

To someone not acquainted with the metamathematical self-reflection of mathematical thought, its systems probably look very artificial and far-fetched. Yet the conceptual motion contained in them is well known to us from experience. We all find in our lives elements of our world, which provide us with an *external identity*, for instance our physical and social surroundings, intimate relationships, friendships, possessions, job, status, nationality, religion, etc. If we are asked in a non-philosophical context: "What are you?", we shall probably answer by mentioning one or more of these external 'characteristics' of our personality. They constitute the externally projected image of our self-identification. But if these determinations change more or less dramatically, as everything in the world changes, we immediately look for something *invariant* with respect to this change, a fixed point, which can serve us as an image of the indifference of our self-identification with respect to its externally projected image. If for instance we lose our job, we identify ourselves by our profession and it is ordained by law that we have a right to a job in the same profession. If we lose a partner, we try to find a more substantial determination of our own personality, e.g. through psychotherapy. If we become estranged from a religious community, we try to find a world-view of our own, etc. We are not satisfied with the indifferent self-consciousness itself, for, as Hegel expresses it, it has no existence (*Dasein*[126]). We always have to identify ourselves with something in the world, a fact which young children manifest clearly by adopting a 'transitional object,' some doll or teddybear which accompanies them continuously for a rather long period. It is well known that its being taken away usually causes considerable psychical damage to the child. The relationship of self-objectification is essential for a subject, and it is this relationship that is externally reconstructed in the metamathematical self-reflection of mathematical thinking. But as in all mathematical reconstruction the original phenomenon becomes hidden by its very reconstruction. The

---

126   See the section on 'self-consciousness as desire' in the *Phenomenology of Spirit*

insight that mathematical construction is *itself* a form of self-objectification of a subject devoting itself to the structuring of the world, escapes mathematical expression.

We may conclude that metamathematics, in its mathematical reconstruction of mathematical reasoning itself, tends towards the philosophical level of reflection, but at the same time remains firmly within the mathematical realm. This tendency is most explicit in the diagonal method, in which a mathematical equivalent of the motion of self-reflection is performed. This equivalent may be summarised as follows:

The subject identifies itself with an objective content in its intentional field. But this identification is more or less arbitrary, and can always be changed. This power of change itself can also be given objectivity. The identification with this objectivity is a relatively unchanging new identification.

Presupposition of this motion of thought is the *arbitrariness* of the identification, in other words: the *externality* of the intentional field with respect to the subject, or the *indifference* of the subject with respect to its objective contents. Therefore this is a typically *mathematical* form of self-reflection.

In the next section we have occasion to recognise this kind of conceptual motion in many reflections of modern philosophers, culminating in Hegel's dialectics. We shall then learn more about its nature, and about its connections to mathematism in philosophy and to real philosophical self-reflection.

3.3 Mathematism in modern philosophy[127]

It is a characteristic of mathematical thinking that it relates itself to something *external* to the subject performing it. That means that it regards the

---

127 In writing this section I have been inspired very much by J.H.A. Hollak's article 'The idea of God in modern metaphysics': De Godsidee in de moderne metaphysica, 1966 pp. 62-74. However, I do not treat modern thought here from the same point of view as Hollak does. He carries out an immanent criticism of the development of modern metaphysics, whereas I try to clarify the relationship of this development to mathematical thinking. In other words, I try to show the attraction mathematical thinking has exerted on metaphysical thought in this period. I presume that this attraction comes from the same historical source as the idea of 'God as causa sui in a positive sense,' which is so central in Hollak's analysis of modernism. This source and its influence on science and philosophy, renaissance-humanism, is investigated very revealingly by E.J. Dijksterhuis in *The Mechanisation of the World Picture*, 1950.

distinctions it creates as indifferent with respect to the unity of its object as well as with respect to its own doings. They are distinctions *in thought only*, such as the imaginary plane dividing the apple into two halves. This apple remains completely the same in my experience, whether thought of as divided or as undivided, and I remain the same in creating this imaginary division. Structural divisions are at the same time no divisions at all, as far as our real experience is concerned. They are only conceptual, not 'existential,' they do not touch or affect us or the world really. Therefore the world of mathematical objectivity is *conceived* of as being purely imaginary. One could not even maintain that it *is* purely imaginary, for that would amount to ascribing to it a nature of its own, while its whole nature consists in being *thought* of in a certain way.

So mathematical thinking is essentially, and therefore necessarily, dualistic, in the precise sense that its subject conceives of its object as completely indifferent with respect to its distinctions. The object, therefore, is conceived of as completely external to the subject of mathematical thought. As a consequence, the object is also completely indifferent with respect to mathematical distinctions, for they are created in thought only, by a subject external to it. These distinctions are conceived of as being hypothetically postulated. What is indifferent to its distinctions is pure *extension*, for something is extended in so far as it can be thought of as divided into parts which have nothing to do with each other, which means again that the whole is thought of as indifferent with respect to the distinctions between its parts.

Descartes describes this relationship of subject and object in mathematical thinking as if it were an *ontology* of real being, in which subject and object - as *res cogitans* and *res extensa* - are themselves thought of as existing indifferently side by side. Not only is the world seen as completely external to the subject, but they are also seen as external to each other, so that a third subject is needed *for which* this externality exists: God, who is understood as *causa sui* in a positive sense (not in the medieval sense of being *uncaused*). God is understood as the foundation of the unity as well as of the distinction of both forms of being. Not only man in his relationship with the world, but also God in His relationship with His creation is understood here according to the paradigm of mathematical thought.[128]

---

[128] Descartes' God is thought of as the *origin* of the two substances and of their concurrence (which is truth). In mathematical thought this origin is the human mind itself. Therefore it is not amazing that eventually Kant makes this God descend from the metaphysical realm into the human mind, which is the origin of the transcendental unity of apperception.

In Descartes' experiment of doubt, the arbitrariness of all opinions, i.e. identifications of the subject with objective contents, is reflected upon. The subject of this *reflection* however, whose indubitable existence is the result of this reflection, is understood as an entity in itself, so in a sense also as objective. Not as an object in the *world*, for that would amount to a form of empiricism,[129] but as an object *outside* the world. The world, as completely outside the subject, and reduced to a field of possible identifications, is naturally understood as *res extensa*. Being an object outside the world, the subject needs another subject in order to be objectified in this way, for it could objectify *itself* only as a subject *within* the world, the world being by definition the field of its possible objectifications. This reflection is expressed in Descartes' proof of the existence of God. The relationship of the human subject to the world is seen as an external objectification of God's self-relatedness as *causa sui*. Therefore God has absolute power over this relationship, and its truth or falsity depends on His will. We (though Descartes did not) understand that it is *mathematical* truth and falsity which depends on such an act of postulating. This is again - but now from a divine point of view - the changeability of self-objectification, the reflection of which results in a necessarily unchanging 'fixed point': the necessary truth of clear and distinct ideas relating the human subject to its world. This necessity, however, is founded in God's will, as the necessity of mathematical truth from *our* point of view is founded in our will to commit ourselves to the logical consequences of the postulated determinations of mathematical objectivity, for instance the axioms of Euclidean geometry. In this respect the God of modern metaphysics was a mathematician from the very beginning.

The relationship of finite and infinite existence in modern philosophy is understood on the basis of the paradigm of objectivity and subjectivity in mathematical thinking. This could not be seen by Descartes and his contemporaries, for the true nature of mathematical thinking had not been revealed yet. Mathematical theorems were understood as eternal truths about nature, not as consequences of postulation on the basis of the principle of structurability. As the specific nature of this principle was not yet recognized, it could only be understood as infinite, and was therefore identified with God's creative power. In this respect Feuerbach and Marx were right: the God of modern metaphysics *is* a projection of a human faculty into the realm of

---

129 This empiricism may be of a psychological nature, where the subject is identified with its real psychical existence, or of a physicalist nature, where the subject is identified with the body as a physical process, or of a sociological nature, where the subject is identified with its place in social or socio-symbolic structures. The psychological variant is found in classical English empiricism, the physicalist form in positivism and the sociological brand in structuralism.

infinity. But this faculty is not the power of sense-perception (Feuerbach) or of changing the world by labour (Marx), but the power to postulate a mathematical universe. They also overlooked that such a projection is necessarily *ambiguous*, for it presupposes the realm of infinity *as such*, which by definition cannot be the result of projection. So while the modern metaphysical conception of God's *nature* may have been anthropomorphic, the idea of an infinite and absolute foundation of all being, truth and goodness was not, and did not differ essentially from scholastic conceptions. This is the truth of Kant's criticism of modern metaphysics.

The difficulty of course was that this idea of *causa sui* was not understood as a *determination* of an already recognized realm of the absolute, but as its very *foundation*. This is precisely the content of the idea of God as causa sui: God's nature is the (formal) cause of His existence. If an existence has a cause, however, it is essentially *finite* in a metaphysical sense, and it presupposes a principle of being, in which the distinction of existence and non-existence is founded. So the distinction of finite and infinite being threatens to disappear in this philosophy. Spinoza had the courage to draw this consequence, and to reduce all finite existence to pure appearance, to modi of the one and only substance, which is infinite and *causa sui*. Spinoza's substance too is essentially ambiguous, for it is the impossible synthesis of finite and infinite being. Yet with its introduction, modern philosophy enters a path which eventually leads away from the mathematical paradigm, although, as we shall see, it will be very hard to escape from its influence. Thought and extension, Descartes' two kinds of substance, are united in Spinoza's substance as two of its infinitely many attributes. This means, that *externality* is *internalized*, and no longer unconsciously presupposed. Because it was originally the unlimited field of possible self-identifications of the subject, pure structurability, it cannot be contained in a finite number of attributes. The relationship of subjectivity and external objectivity explodes into an infinite firmament of possible identities of the absolute substance, the infinitely many attributes.

As this numeric infinity still embodies the idea of externality, its consequence is Leibniz' philosophy, in which this inward explosion is countered by an outward explosion of an infinity of monads. The idea of relationship, confined within the abstract idea of substance, strives to break out. In Kantianism and German idealism it gets the chance to do so explicitly, but not with complete success. Even Hegel's philosophy does not get rid of Spinoza's ambiguity of finite and infinite being. The *real* relationship (not merely the relationship as it functions in mathematical thought) of human subjectivity and objectivity is developed adequately, but it is not understood as belonging essentially to a finite way of being. The human spirit is *causa sui* in the sense that it meaningfully structures its objective world of experience, and so determines itself as the subject of this world. But in order to do this *meaningfully*, it has to be *receptive* with respect to cues indicating meaning. Hegel, indeed, has *exerted* this receptiveness in a genial manner, but he has not given it the proper place in his philosophy. It is connected to what he calls 'finite knowledge,' i.e. knowledge in which the known remains external to the knowing subject. For absolute knowledge there is no 'outside,' so receptivity is considered to be irrelevant here. And as there can be no finite knowledge for an infinite subject, the human subject must be understood as a self-objectification of this infinite subject, somewhat as it is in Descartes' philosophy.

This position leads to a systematic ambiguity in Hegel's philosophy, for it becomes impossible to determine adequately the role of externality itself, and of the relationship of finite and infinite being. The ambiguity concerning externality becomes almost explicit in the transition from the 'Science of Logic' to the 'Philosophy of Nature,' in which the absolute Idea is said to 'liberate' itself into external existence, which is nature. This can be interpreted as an act of creation of the world, and it has been criticised as such by Schelling. For how can one understand that a purely logical entity creates the world? On the other hand, one can understand the 'Science of Logic' as expressing the transcendental conditions of the possibility of the *human* spirit, which *presupposes* the existence of this spirit and its world of experience. The 'self-liberation' of the absolute idea then means that it turns itself from the pure self-reflection of the conditions of its transcendental openness towards the actual exertion of this openness with respect to his world of experience. Hegel gives cues for both interpretations, which may mean that he did not really see a distinction between them.[130] But this ambiguity permeates all of Hegel's philosophy, for externality is a necessary ingredient of the relationships in the realm of the *human* spirit, but not of relationships dependent on the notion of spirit *as such*. As Hegel is not clear about the essential distinction between the two, it is obvious that the role of externality must always be ambiguous.[131]

The role of the relationship of finite and infinite being is affected by a similar ambiguity. This is most clear in the 'Lectures on the Philosophy of Religion,' in which it is almost explicitly expressed in the adagium that without a world, God is not God. This may be interpreted in a pan-entheistic way, in which the creation of the world is reduced to an immanent self-relationship of infinite being. But it may also be interpreted as an expression of the finitude of the human spirit, which cannot have any knowledge of infinite being otherwise than by reflection on the experience of the finiteness of its own existence. This ambiguity appears in Hegel's writings as soon as he has to deal with the relationship of finite and infinite being.[132]

---

130 Cf. L.E. Fleischhacker, Gibt es etwas ausser der Äusserlichkeit, 1990.

131 Cf. L.E. Fleischhacker, Hegels Entwicklung der logischen Prinzipien in der Wissenschaft der Logik, 1991, pp. 83-98

132 Cf. J.H.A. Hollak, Wijsgerige Reflecties over de Scheppingsidee; St. Thomas, Hegel en de Grieken, 1975, pp. 89-104, and also my article mentioned in note 131.

Human existence as finite self-manifestation and finite self-determination is in Hegel's philosophy projected indeed 'ideologically' into the realm of infinite being. This has been grasped rightly by Marx and Kierkegaard. But the other side of the story is that Hegel's whole philosophy is ambiguous, in the sense that finite and infinite being cannot be distinguished adequately at all in it, for his notion of infinity - external unboundedness - equally fits the human spirit and the absolute spirit. The human soul is προς 'ολον 'in a sense everything,' as Aristotle already knew,[133] and which is affirmed by philosophical anthropology in our time. Yet it is finite in the sense that it is confined to the human condition. As Sartre expresses it: it is condemned to freedom. It cannot exist in any other way than by determining itself through the process of meaningfully structuring its objective world or, as J. Hollak formulates it, it has only a *virtual* intuition of itself.

We cannot get rid of the responsibility for the meaning we give to our existence, and live like an animal, nor can we, like a pure spirit, acquire clear intuitive knowledge of ourselves independently of an objective world; even less are we the origin of the fact that we and our world exist. We are finite bodily spirits, a fact which is differently expressed by different philosophical anthropologists, but always in one way or another affirmed. Now here we are back at the point where Descartes proof of the existence of God started. The human spirit knows that it has a bodily - and therefore *historical* - and finite nature, and is forced thereby to recognize the existence of infinite being as the cause of its own existence. But this does not imply that it can understand the *nature* of infinite being or even reconstruct its necessary doings. Nor does it follow that the relationship of the infinite cause to its effect is comparable to the relationship of subjectivity and objectivity in the realm of human spirit. On the other hand we cannot convincingly succeed in making a sceptical or nihilistic dogma out of this non sequitur. For the precise boundaries of the range of our intellect are not given either. Here our thinking finds itself liberated to enter the open space of metaphysics, which is "*As wide as all reality.*"[134]

We have walked in this section through modern philosophy with 'seven-league boots,' starting with Descartes' philosophical counterpart - and probably the original paradigm - of the mathematical diagonal method. We have tried

---

133 In fact Heraklitos already expressed a similar feeling: "You'll not be able to go and discover the limits of the soul, whatever road you may go: so inexhaustible is what it has to say." (fragment 22B45)

134 Title of a book by the Dutch metaphysician Herman Berger: *Zo wijd als alle werkelijkheid*, 1977.

to follow the gradual elimination of his mathematism, which consists in always falling back into an external reflection on results. We have seen that Hegel almost succeeded in eliminating the last remnants of this attitude. We have not yet seen, however, the situation of post-Hegelian philosophy, which usually rejects the good with the bad, and therefore still lives with the mathematistic heritage of modern philosophy. I shall pay attention to the contemporary forms of this heritage in Chapter 5, when dealing with the anti-metaphysical attitudes appearing in the twentieth century.

This chapter will be concluded with an approach towards a true philosophical counterpart of the diagonal method, which involves the attempt to distinguish its philosophical from its mathematical side.

Characteristic for the mathematical context of the method is the externality of objectivity with respect to the subject. The action of the subject can only be *projected* into objectivity or mirrored in it. *As* an action of the subject it has no locality, no coordinates[135] in mathematical objectivity. It can *give* itself coordinates, but not as real action, only as an intentional counterpart of it, as *representing* the action. In information technology this representation is embodied in the *physical* action of the hardware system, so that we are now looking into a *material* mirror, not only into the ideal mirror of mathematical objectivity. But an information-processing system remains a mirror-image, experienced as an *external* reflection, which may be useful and fascinating, but also indifferent to us. It can be changed without really affecting our world of experience. Of course the *effects* of the change can eventually affect our world, and are even *used* in technology precisely for such a purpose.

The indirect, external self-awareness of the subject in the form of its external self-objectification presupposes a *direct* self-awareness as a necessary component of all intentional acts. In scholasticism this was called the *second intention*. We cannot understand anything without somehow understanding *that* we understand, and we cannot perform an act of willing without somehow willing to perform it.[136]

---

135 I use the mathematical image of coordinates, i.e. numbers determining a place in space, consistently following the idea of projection. One imagines a space of higher dimension to be projected into a space of lower dimension. What is wrong with this image is clear: the subject as such is not to be localized in *any* space at all.

136 Kant denies, at least in the realm of theoretical reason this immediate self-reflectivity of the subject. He distributes spontaneity and receptivity over the faculties of intellect and sense-perception. So for the intellect there is only spontaneity, and no direct awareness of anything else than what is given by sense experience. Not only the noumenal world

The difference between this *inner* reflectivity of our mental acts and the *external* reflection we find in the diagonal argument can be illustrated by means of the so called *liar's paradox*. The classical form of it is well known as the story about the Cretan who proclaims that 'All Cretans are liars.' This form has a 'way out' consisting of the situation in which not all Cretans are habitual liars, but our particular Cretan surely tells a lie if he tells us they are. The modern form of the paradox can be illustrated by the sentence: 'This sentence is not true', understood in a self-referring way. It is claimed that this gives a contradiction, for if the sentence is true, it must be false, so by indirect proof it is false, so it must be true, but that is a contradiction. Now one may distrust the self-referring character of the sentence. One might try to make it more precise in three ways.[137]

First, one could make the sentence refer to itself as an *inscription*, marking the place where it is to be found, e.g. by saying 'The only sentence in the book *Beyond Structure* by L.E. Fleischhacker containing the sign @ is false.' (Let us name this sentence A) But then the identity and therefore the truth or falsity of the sentence would depend on its place. This means that it is not permitted to make the transition from a statement of the form 'Sentence A is true' or 'Sentence A is false' to the conclusion that what sentence A *says* is true or false. So the inference: 'A is proven to be false, but it *says* of itself that it is false, so it must be true.' fails in this case. This is illustrated by the thought experiment of designing a computer program for checking sentences on the basis of their inscription (which means their address, e.g. in the computer memory). Such a program would necessarily be led into a loop by sentences such as A, because it would have no means to *interpret* the sentence. There is no contradiction in this case, only infinite regress.

The **second** way to specify the self-reference is mathematical. We give all words in e.g. the Concise Oxford Dictionary a number, and give a rule of composition so as to calculate a unique number for every sentence from the numbers of its words.[138] Moreover we give a number to each decimal

---

of things-in-themselves therefore remains inaccessible to the intellect, but also its own activities, which in their actuality also must belong to this world.

137 See A. Koyré, *Epiménide le menteur* Paris s.d.

138 This is essentially the method of Gödel-numbering.

numerical expression,[139] and on that basis to each arithmetical expression formed from these, and add these expressions as 'words' to the system.[140] Now it is possible to calculate the number of a sentence, which is e.g. obtained from another sentence by replacing the fifth word in it by another word, from the number of the original sentence and the number of the replacing word. Say this calculation method is expressed by an arithmetical expression, which we indicate by S. So $S(m,n)$ indicates the number of the resulting sentence after replacement of the fifth word in the sentence with number m by the word with number n. Now let the sentence 'The sentence with number unknown is false.' have number g. Then the sentence 'The sentence with number $S(g,g)$ is false.' exactly has number $S(g,g)$, so the self-reference has succeeded. The contradiction we obtain now disproves the silent hypothesis that truth and falsehood are determined in an unequivocal way by the syntactic composition of the sentence. This is essentially Tarski's theorem referred to in section 3.2.

The **third** way of self-reference is based on the inner reflectivity of mental acts. It is precisely the way in which Descartes established the necessity of his cogito by considering that in order to think, one must be.[141] By reflecting on the act of thinking while doing the very thinking which includes this reflection. One might try to understand the sentence: 'This sentence is not true.' in the same way. The contradiction now disproves the silent hypothesis that the sentence has the definite meaning of a *judgement* at all.

Now it has already been said that the **first** method does not involve a contradiction. In fact it leads to an endless repetition of the motion of self-reference without noticing that it *is* a motion of self-reference.

The **second** method does not lead to a categorical contradiction, but to a *conditional* one. The condition is that the predicate of truth for sentences is definable on the basis of their numbers, within the given mathematical context. So it follows that this condition is not fulfilled, which is Alfred Tarski's result on undefinability. It is a typical application of the genuine diagonal method, leading us outside a given mathematical context.

---

139   Such as '2385674.' The number of the expression as a syntactical object is usually not the same as the number indicated by the expression.

140   That all this is possible must be, and in fact *has been* shown mathematically.

141   Mais, aussitôt après, je pris garde que, pendant que je voulais ainsi penser que tout était faux, il fallait nécessairement que moi, qui le pensais, fusse quelque chose. *Discours de la méthode* quatrième partie, *Oevres Philosophiques*, Edition Alquié, Tome I, p.603.

The **third** method also leads to a conditional contradiction. The presupposition here is that the logical subject: 'This sentence' refers to a real subject, the content of a mental act: a judgement. In the 'cogito' the actual thinking is such a content for *I am really thinking*: while thinking I know that I am thinking. But in the liar's case a contradiction would result if there were such a content, which simply proves that there *can be no content*, the logical subject cannot refer to a real subject. The ambiguity of the word 'sentence' *suggests* a real subject: either an inscription or a grammatical structure, but these interpretations fail to produce an unconditional contradiction. The interpretation, however, that *should* give the contradiction is immediately frustrated by the form of the sentence. It is very well known that not every grammatically correct sentence expresses a judgement. The liar's sentence in this form is so designed that it *cannot* express a judgement.

Both the mathematical and the philosophical interpretation of self-reference lead to the negation of a silent presupposition here. But in the mathematical case we can always introduce more comprehensive systems and new silent presuppositions. We cross only a contingent border of some more or less arbitrary system. In the philosophical interpretation, however, we hit upon an absolute impossibility, or, in Descartes' case, an absolute necessity.

Such a result is of a curious nature, for it cannot be *explained* by anything else than the act of reflection itself. If one tries to explain it, it is either reduced to the mathematical case, in which the necessity of the result becomes only relative with respect to a certain context, or to the (quasi-)empirical case, in which the necessity disappears altogether. This is the *speculative* character of philosophical reflection. Just like sense perception, the other extreme of the spectrum of our epistemic faculties, it has a 'take it or leave it' character. That character seems to make it very vulnerable to positivistic criticism. Now for positivism of an *empiricist* nature, the philosopher can point out that the basis of such an attitude involves the same problem. As an effect of such criticism positivism has always tried to find a handhold in mathematics. Now it has become clear that a *particular* mathematical theory can only provide such a handhold if one adheres to its axioms dogmatically.[142] So one is forced to direct one's attention to mathematics *as such*, or in general, to the mathematical *method*. As this method cannot in its turn be based on some particular theory, its trustworthiness and evidence must be based on a *principle*, which can only be known *explicitly* by philosophical reflection: the very method the positivist

---

142   Which, in fact, is very unmathematical.

wants to combat. The last escape seems to be the denial that the mathematical method has any positive content. This means that it is only regarded as a *language*, in which the facts of science may be expressed unambiguously. But then the whole rigmarole starts anew, for what then is the basis of knowledge of these 'facts of science'? Now the choice is again between two ultimately unexplainable forms of cognition: perception and reflection.

Probably as part of a strategy of defence against positivistic attacks, the necessity of the results of philosophical reflection is sometimes explained by the argument that 'the contrary is impossible (or even inconceivable).' Of course it is true that the contrary of a logically necessary state of affairs is impossible, but is that an explanation? The contradictory character of the statement of the contrary has no other content than the apodeictic character of the statement of the state of affairs itself. So it is an explanation of 'the same by the same.'

If not *logical* but *real* impossibility of the contrary is meant, there is no difference between factual and essential states of affairs, because *after an event has taken place*, the coherence of its characteristics is as necessary as it is in the case of an essential inner coherence of the kind philosophical reflection tries to grasp. The fact that a certain event really has taken place can of course be denied afterwards, but these denials will be false for ever if this fact indeed is a fact. The only difference with respect to essential coherences is that the falsehood of the denial of the latter has no beginning.[143]

Such a defence 'from the negative' of the fundamental and necessary character of the results of philosophical reflection is impossible and superfluous.[144] Like a scientific experiment and a mathematical proof, a

---

143 Traditionally a distinction is made between absolute and conditional necessity, the first being ascribed to states of affairs with an a priori character, the second to factual states of affairs. It should be noticed that 'conditional' does not refer to a 'strict implication' here (It is necessary that if A then B) but to a statement of the form: 'If A, then B is necessary.' The classical example being: 'If Socrates is seated, then he is seated necessarily' Therefore the conditional necessity is not a mere case of the non-contradiction principle. One should also notice that epistemological questions have nothing to do with the ontological distinction between the factual and the a priori. We can err in our presumed knowledge of the a priori as well as we can err with respect to facts.

144 It does not provide any methodological advantage either, for it is only the content of the essential coherence which can tell us anything about its essential nature. This can be compared to the situation in mathematics. Although there is a formal criterion for inconsistency of a theory, there is no effective formal criterion for consistency. Only the

philosophical reflection is unrestrictedly repeatable by anyone having the necessary training and knowledge to do so. And like these two methods of unquestioned reputation, it may lead to false results, correctible only by using the same method again, while avoiding some former mistake. What more epistemological status do we want? Our knowledge of the necessary is by no means necessary knowledge. We may err there as anywhere else, and we need not be ashamed of it, as the history of philosophy proves. No additional philosophical credo, such as 'fallibilism' is necessary for that. It is even dangerous to insist on fallibility as an explicit philosophical principle, for if it is formulated explicitly in a certain form, a dogmatic surplus may be smuggled in by the form, concealed by the original evidence of the content. On the other hand we have to admit that such a danger is inherent in any explicit formulation of principles.

It has now become clear what has gone wrong in modern metaphysics. The paradigm of the mathematical approach in the natural sciences has seduced the great modern philosophers to try to imitate mathematical thought. This has led them into an indefensible position, which has been unmasked by Kant. The German idealists tried to repair the defect, but even they were influenced by the mathematical paradigm. Hegel, for instance, regarded his dialectical method as a kind of αποδειξις, a kind of *proof*. Speculative philosophy, however, is not strengthened by allowing mathematical thought to dominate it. On the contrary, philosophical exactness is of such another nature than mathematical exactness, that any mix of them becomes turbid. The hard labour of separating again the two forms of thought, initiated by Edmund Husserl,[145] is still in full progress.

---

content of the theory can give us any trust that it is consistent. If we do not succeed in denying a certain connection, e.g. between unity and plurality, it is because we are already convinced of its essential nature.

145 Husserl's *Logical Investigations* can be seen as a search for proper philosophical accuracy without doing injustice to mathematical exactness, which was very much admired by him. In this respect I regard the present work as an attempt to bring this - as yet uncompleted - attempt a step forward.

Chapter 4
The degrees of reflection: systematics

After having obtained some experience with the ambiguities and shifts of perspective involved in the distinction of three levels of reflection, the reader will now probably feel the need for some systematization. Such a systematization requires a philosophical explanation of the concept of *degree* as it is used here. We have already seen that such terms as 'level' and 'dimension' suggest a non-philosophical discourse. An attempt at a dialectical development of the degrees, which would characterize them as 'moments,' will be seen to give rise to difficulties of quite another nature. It does not provide a proper basis for the idea of *gradation* which is inherent in the original scholastic paradigm - the degrees of abstraction -, and which provides the present approach with the elegancy and clearness constituting the value of the whole enterprise. The distinction of degrees of reflection forms a philosophical problem in itself, ultimately leading to the metaphysical considerations of the next chapter.

4.1 Levels, dimensions and moments

The picture that comes to my mind when I think about the degrees of reflection is that of a sea, representing the totality of experience. If we were merely immersed in it, without any protection, we would drown. Normally we are sailing on it by means of the boat of common sense, along with a community of fellow-passengers. The vessel becomes our world, but the sea bears it and rocks it. We cannot look deep into the water, but we see its surface and we are content with that.

When we fly in an aeroplane above the sea, we are independent of its motions. The other passengers are our company only for a short time. We look deeper into the water, but we have a less detailed view of the qualities of its surface, though by measurement we can reconstruct many details. Our perspective is determined by the height at which we fly and sometimes clouds blur our vision. But by measurement and reasoning we can discover many interesting facts, unknown to the boat-passenger.

From the space-shuttle we see the globe of the earth, the basis of our very existence. We realize that the firm ground of the earth is not so very different from a ship. The earth is floating in space just like our shuttle. The reality of our earthly life is sublated[146] and therefore vigorously confirmed as being *our* reality. But at the same time we are wondering what this reality is, where it comes from and what will become of it.

This allegory of the three degrees of reflection mixes up various elements of our experience: practical, emotional, geometrical, existential. It can be said literally to describe three levels. It can also be said to describes three dimensions of experience, say quotidian, alienated and existential experience. Words such as 'levels' or 'dimensions' may be unphilosophical if understood literally, yet in a metaphorical sense they are not inadequate at all.

When it comes to philosophical precision, however, metaphor is not enough. We have to investigate which of the elements mixed together in the metaphor contains the idea we are looking for.

What seems to be adequate in the concept of *level* is the idea of increasing distance. But in such a way, that increasing distance does not necessarily imply decreasing contact. In physics interaction usually decreases with increasing distance, but, as the quarks show us: not always. Mathematically, contact cannot be defined otherwise than by contiguity, so in that sense *any* distance completely excludes contact, but physical contact means interaction, which is of quite another order. What we really need here, of course, is the idea of increasing distance which is at the same time increasing contact in a certain specific sense. This is no contradiction in terms, provided we do not take the mathematical discourse as our paradigm. By taking more distance, we see the world with other eyes, we see less detail, but more global shape and more context. We are less involved with the short waves of life and more with the long waves. We need not be less emotionally involved. The superficial emotions even out and the deeper emotions become conscious.

What is most misleading about the *level*-terminology, is that it contains the idea of arbitrariness. There can be as many levels as one wants. It only depends on the accuracy of one's measuring method. So why three? Distance can be changed continuously, so any partition is arbitrary. In that respect the

---

146 This word, which might be absent in most dictionaries, is invented by J.R. Findlay as a translation of the Hegelian term '*aufgehoben*', which has a triple meaning in German: a) put out of our way; b) preserved; c) moved to a higher level. All three meanings are present when something is put on the highest shelf. Findlay J.N., *Hegel; a reexamination*, 1958.

idea of *dimensions* seems to be more attractive. If we avoid the kind of definition which leads to 'fractal dimensions' and keep to the usual definition, dimensions are natural numbers, and although there is no *mathematical* reason to give three-dimensional space a special meaning, its direct applicability to the space in which we move our bodies yet gives it such a meaning for common sense. Dimensions are present at any point in space, which is in accordance with the idea of all three degrees of reflection being involved in any knowledge. Dimensions, contrary to levels, are not separable from space, just as the three degrees of reflection are not separable from the totality of experience. Only the idea of increasing distance is lost.

Hegel derives his use of the word *moment* directly from its etymology: *movimentum*: that which moves, in the sense that it sets something in motion. This leads to a difference between his use of the word and the usual use of it in mechanics in the sense of momentum or of torque. In mechanics a *product* of two factors determining or causing motion corresponds to this concept. Mass and velocity, together determining impulse; force and arm, together determining torque. In Hegel's use a moments should be compared to a *factor* in the product, not to the product itself. It is inseparable from the concrete phenomenon and moments are determined only relatively with respect to each other. Factors may change without any real consequence if their product - representing a real effect - remains constant. Only *together* have they any effect. So, for instance, subjectivity and objectivity are inseparable, relatively opposed and necessarily present in any reality. Contents of experience may indifferently be understood as objective or as subjective. It is only the *relationship* of subjectivity and objectivity that has anything to do with concrete reality. In the same way one can think about unity and plurality, identity and distinction, and so on. In dialectical thought, such a relationship of moments can be expressed in three stages. *At first* their totality is present in an undifferentiated manner, as it is in a concrete phenomenon appearing to us at first sight. *Secondly* the opposition of the moments can be elaborated, so that it becomes clear that they exclude each other relatively, and their immediate identification leads to a contradiction. In the *third* stage, it is possible to show that their relationship as such is their proper reality, which we now realize to have understood already implicitly in the first stage.

Can one think about the degrees of reflection as of such a dialectical development of a relationship of moments? The moments may be named *experience* and *understanding* and in the *first degree* of reflection they are undistinguished: we take our experience to be identical with what we understand of it, with what we can express for ourselves in concepts, and for others in language. There seems to be no opposition between being and knowing. In the *second degree* of reflection there is a clear opposition of two worlds: the ideal world of structure and the real world of experience. Whatever models of our experience we construct, they always remain models, and never become reality. The contents of experience may behave in a very 'structural' way, yet they remain qualitative, and can never be reduced to pure structure. Pure structure, on the other hand, can become infinitely complicated, but will remain incapable of any action whatsoever. The *third degree* of reflection reminds us of the relational character of the connection between model and experience, as well as of the connection between concept and reality. We see that these are correlatives which we must try to understand from their relationship.[147] To consider them as immediately identical is naive, to consider them as completely separated is artificial, and therefore equally naive.

Such a dialectical approach has many advantages. It does not depend upon metaphors or other imagery, but has a coherent conceptual basis. It shows the relationships between the degrees and explains their differences at the same time. It clarifies the degrees as degrees of self-understanding and it makes us feel the philosophical naivety of the first and second degree.

Yet there is a problem. For, however satisfying this point of view may be to the philosophical mind, we might feel that it does not do complete justice to common sense and mathematical thought. For these seem to be understood mainly as *incomplete forms of philosophy* and not as perspectives in their own right. I am convinced that the opposition against Hegelian philosophy in his own and in our century has something to do with this. It is perfectly true that we cannot understand the third degree of reflection as being based on abstraction in the sense of leaving out certain aspects of experience, like the first degree leaves out the concrete individual and the second leaves out quality and change. But it is equally true that we cannot live in a completely philosophical universe, without common sense and mathematical construction. These two forms of abstraction have their own epistemological right. But how is it possible from a philosophical point of view, to do justice to non-philosophical understanding? On the other hand, if it fails to do so, it takes the risk of placing itself outside the field of common human interest.

---

[147] This is precisely the project Hegel undertook in writing his *Phenomenology of Spirit*.

4.2 Philosophy and experience

The human spirit has often been characterized as coming into contact with itself only through its opposite. It lives in the objectified material world, identifying itself with the assimilated contents of its experience and it realizes itself in this world by practical action, knowing itself mainly by the results of such action. Philosophical self-knowledge is without doubt a necessary component of this process, but only under the condition that it is itself a 'moment' of this process. We may compare this state of affairs to ordinary practice. In practical action there is always and necessarily an element of reflection, but if reflection becomes dominant it hampers action. So in human life, there is necessarily philosophical reflection, but: *primum vivere, deinde philosophari*: living comes first, and philosophizing follows. Reflection of the first and second degree is more involved in practical life than philosophical reflection. The common-sense world is the world we live in, and the world of structure is involved in all technical forms of control. The epistemological relation between the mathematical degree and the philosophical degree has already been investigated in Chapter 1. Although the principle of structurability, which is the foundation of mathematical thought, can be understood properly only in a philosophical way, the unfolding of the possibilities enclosed in it can only take place by actual structuring, either in an ideal way in mathematics, or practically in technology. The latter activity presupposes an implicit understanding of the principle. On the other hand the philosophical understanding presupposes the experience of this unfolding.

Now this state of affairs does not yet contradict the dialectical development of the degrees of reflection in section 4.1, for it is an essential element of such a development that all stages presuppose each other in their own specific way. It is only in the epistemological *evaluation* of the stages that the problem appears. For what is it we can *know* by mathematics that we cannot know by philosophy? Not the principle of structurability itself, for we know that only implicitly by mathematics and explicitly by philosophy. What we know more in mathematics is *how we can unfold the principle* into a world of mathematical structures and technical constructions. That is to say, we know the principle *as it is embedded in the totality of our experience*, whereas philosophically we only know *that* it is so embedded. Analogously in the first degree of reflection we know *how* our world of experience can be conceptualized, philosophically we only know *that* it can be conceptualized. We cannot philosophically deduce a particular conceptualization, nor can we deduce a particular mathematical theory or technical construction. In more Aristotelian terms: the principles are causes of real potencies, but knowledge of the principles is not sufficient to actualize these potencies, neither in reality nor in thought. In order to actualize the potencies we need an immediate contact with experience, which cannot be produced by philosophical reflection, for it only understands the intelligible principles of it, not the particular possibilities of their unfolding.

By thus formulating the relationship of the degrees, we leave the dialectical universe of discourse behind us. The degrees are now not only distinguished by a development from implicit to explicit knowledge, but also by a development from a more vital to a more intellectual contact with our experience.[148] This is probably what the 'vitalist and existentialist' critics[149] of German idealism must have felt when they tried to put Hegel 'from his head onto his feet.' But what they overlooked was, that whatever they took for vital or existential reality was a product of reflection as well. With respect to the role of mathematical reflection this also holds for the more 'positivist' criticisms of idealism. It has been a slow process of discovery of the philosophy of science of our century that the 'scientific facts' the positivists opposed to idealistic notions were produced throughout by a process of mathematical structuring of the world of experience. For the first degree this work has been done by the structuralists, who made it clear to us that the so called 'immediate life-world' is a cultural product closely connected to language. Of course, now there is a tendency to draw nihilistic conclusions from this justified criticism of idealism instead of understanding it as an opportunity for the rehabilitation of philosophical reflection.

Finally the classical distinction of the degrees of abstraction has to be reconsidered. It has already been explained extensively in Chapter 2. The first degree gives us general knowledge of material things as such. That is to say that the *content* as well as the *real object* of such knowledge is of a material nature. The second degree gives us knowledge of material things in an intelligible form. The real object is still material, but the content is its intelligible structure. The third degree gives knowledge of what is immaterial both in content and existence. This does not imply that it cannot be *given* to us by

---

148 Perhaps, we can find this ambiguity in all dialectical developments. Concerning Hegel it may even be maintained that there is no ambiguity, and that the vital side is fully recognized. Yet there is a tendency in his philosophy to reduce it to the 'natural,' which is essentially the 'external,' a category which he fails in my opinion to determine positively. Cf. L.E. Fleischhacker, Gibt es etwas ausser der Äusserlichkeit? Über die Bedeutung der Veräusserlichung der Idee, 1990, pp. 35-41

149 Under the denominator 'vitalist and existentialist' I sweep together here such diverse thinkers as Kiergkegaard, Marx, Nietsche, Klages, Bergson, Scheler and Sartre. They all have the intuition in common that there is more in 'real life,' or 'human existence' than the idealists have dreamt of in their philosophies.

experience in the material world. Therefore I interpreted this knowledge as knowledge of the immaterial *principles* of the material world.

This view can be completed for the conception of degrees of *reflection* by including the forms in which these kinds of knowledge are expressed and elaborated: the conceptualized world of quotidian experience; the world of science and technology and the world of genuine human wisdom and practice.

The classical view has the advantage that it avoids the problem of the dialectical view: it includes complete recognition of the lower degrees as producing knowledge of an object not accessible to reflection of higher degrees. On the other hand it leaves us at a loss concerning the question what the relationship may be between what we have called the 'immaterial' and the 'material,' in content as well as in mode of existence.

In order to be able to express the idea of the degrees of reflection systematically, obviously I have to find a synthesis between the dialectical and the classical view. For that purpose I cannot avoid specifying what status I ascribe to human knowledge in general.

4.3 Anthropological considerations

Recognizing that knowledge of principles presupposes a totality of experience, potentially containing more than this kind of knowledge alone, is nothing more than reflecting on the concept of principle as such. But this whole exercise, to be meaningful, presupposes that the human condition allows of these kinds of knowledge, that is, that reflection of each of the degrees corresponds to some sphere of reality contained in the human condition of existence.

In Chapter 2 it has already been explained what these spheres are. The first is the sphere of common sense and 'life-world,' the second contains at least technology in the broadest sense of the word, and the third has to do with the 'existential' element of human experience.

It is also possible to interpret these spheres as corresponding to degrees of *excentricity* in Hellmut Plessner's sense[150]. This could be elaborated in the dialectical manner of section 4.2. The first degree then corresponds to a form of human life in which the excentricity - the discrepancy between immediate self-experience and reflection - is lived, but not recognized. In the second degree it is not only recognized, but actually *used* for practical

---

150 See e.g. *Conditio Humana* or *Philosophische Antropologie*. The notion of excentricity is so fundamental in those works that is explained extensively right at the beginning.

purposes, however, without questioning the possibility of switching between the immediate and the reflective points of view. In the third degree this question is central and human life is then actually *experienced* in its excentricity. As has been noticed in section 4.2, this gradation must also be seen as a development from a more vital attitude in life to a more intellectual one. This does not mean that in the first degree the intellect is absent, nor that in the third vitality is absent. It is a question of a shift of accent. In the sphere of common sense we have to use concepts and to judge situations without hesitation. To do otherwise would often be dangerous. Our judgement may be wrong, and we know it, but we cannot ponder too long about it. Better to make a mistake than not to react at all, for that is seldom adequate. In a mathematical or technical attitude we take greater distance. The force of technology is the possibility of a roundabout way of doing things, which is incomparably more efficient. And that is precisely the power of mathematical thought too. We are looking here for a certain equilibrium between vital need and an intellectual appreciation of the situation. In the philosophical attitude we can postpone our judgement and try to experience the heart of the matter. It has often been said that this too contains a striving for power, as though power as such were a bad thing. Certainly, the human spirit cannot do without power, that is, without the capacity to realize itself in the world of its experience. For that is its way of existing. As a pure spirit it could not exist at all,[151] its intuitive force is too weak for that. It has to express itself in its world in order to know itself, which is to exist as a spirit. Therefore it can only exist as an embodied, that is *vital* spirit, endowed with a striving for power, which does not mean that it has to live at the cost of other beings. Power is clearly distinct from violence, which means robbing some being from its faculties to exist. So also in the philosophical attitude there is vitality, and wisdom is a power in itself, but the kind of power which is the least violent of all. In order to realise itself as this power, it must free itself from the hypocrisy of conceiving of itself as powerless.

We now have another clue for the number of degrees: two extremes and a middle. Vital striving may be dominant, intellectual striving may be dominant, or the striving after an equilibrium may be dominant[152]. At first

---

151 Whether the notion of an unembodied finite spirit makes sense at all, e.g. in the way medieval scholastics tried to understand the existence of angels, is in my opinion still an undecided question. Only if this question can be answered positively, questions concerning the factual existence and ways of manifestation of such spiritual beings can be asked meaningfully.

152 This is how J. Berghuys understands the position of mathematical abstraction between common sense and philosophical thought in his *Grondslagen der aanschouwelijke*

sight this seems to have nothing to do with the epistemological distinction between conceptualizability, structurability and intelligibility of principles. Yet we remember Aristotle's characterization of mathematical objectivity: *'the sensible, but not as sensible.'* The objectivity corresponding to mathematical reflection is characterized by a certain equilibrium between the intelligible and the sensible. It is 'intelligible matter,' the materiality *as such* of material being, that is responsible for structurability. In Chapter 1 it has already been noticed that this equilibrium between sensibility and intelligibility is precisely what constitutes mathematical 'exactness.' We can therefore characterize mathematism as a kind of undue modesty of the intellect. The highest form of understanding we can strive after, is seems to say, is the form in which it is in equilibrium with perception. The paradox of this kind of wisdom, if it is wisdom, is that it surpasses its own ideal precisely because it *postulates* this ideal as an ideal. Such a postulation cannot be justified by mathematical reflection itself. It needs philosophical reflection, which it rejects.

4.4 The philosophical point of view

If we now try to look at the problem exclusively from a philosophical point of view, what remains is only the distinction of three epistemic principles: conceptualizability, structurability and intelligibility. These principles are expressed as potencies, for they have been derived from the corresponding activities, reflection as it comes in three degrees. What we mean by a principle, however, is not a potency as such, but the *cause* of a potency. And as the potencies are here potencies of understanding, their causes must lie in that which is understood in relationship to that which actually understands. Here we have returned to Plato's problem of determining the ontological status of material, mathematical and ideal being on the one hand, and to Aristotle's problem of expressing the activity of the intellect on the other. There is an ontological as well as an epistemological question.

Epistemologically we can *almost* be satisfied with the dialectical explanation of the degrees, but does this answer the ontological question?

The traditional Aristotelian-Thomistic answer to this ontological question is, that some objects of knowledge are dependent upon matter in their mode of being and some are not. Of those that are dependent, some also contain matter as part of their essence and others do not. What is dependent upon matter in its mode of being and also in its essence, is an object of knowledge

---

meetkunde, 1952.

in the first degree of abstraction. What is dependent upon matter in its mode of being, but not in its essence, is an object of knowledge in the second degree of abstraction. What is not dependent upon matter in its mode of being at all, is an object of knowledge in the third degree of abstraction. We can easily recognize: 'material substance,' 'structure' and 'principle' here. We cannot, however, be satisfied completely any longer with this kind of explanation. It can be interpreted too easily as being based upon an *objectivistic* metaphysical perspective, as if material substance and mathematical structure were totally independent of our activities of conceptualizing and structuring the world of experience.

An alternative to such a perspective is to be found in the most elaborated form in Hegel's philosophy. What account would he give of our three degrees of objectivity?

In the science of logic the first part is called 'being.' It deals with what is immediately experienced as objective reality. It is divided into three parts: quality, quantity and measure. In quality being is immediately identical with its determinations. In quantity it is indifferent with respect to its determinations. In the chapter on measure the relationship of immediate identity and indifference is worked out. Now in quality and quantity we can recognize the formal objects of our first and second degrees of reflection. Measure, however, unites the two; and reflection on principles as such starts only in the next part of the science of logic, which deals with essence.

It is clear that there is no easy mapping to be found between Hegel's logic and the classical theory of the degrees of abstraction, which could be the basis of a synthesis. Therefore it is necessary to make a new start, keeping in mind, however, these two approaches. The epistemological and the ontological side should, in this approach, be in a certain precise sense *independent* of each other. A form of knowledge can be completely adequate with respect to a certain form of being, but not completely *in*adequate with respect to others. Mathematical knowledge is indeed completely adequate with respect to what is thought of as mathematical objectivity, but not completely inadequate with respect to ordinary material being. On the other hand conceptual knowledge is not completely inadequate for mathematical objectivity, although the ideal is a complete structural formalisation of mathematical concepts. In philosophy we can do neither without concepts nor without structures, although we constantly strive to surpass them. My attempt at a philosophical conception of the degrees of reflection is inspired by the classical theory on the ontological side and by dialectics on the epistemological side, and, when considering their connection, by Max Scheler's conception of congeniality between forms of knowledge and spheres of being.

a) The ontological side

In reflection of any degree, experience is in one way or another understood as *objective*, which means that some *form* of being, which is distinct from its purely intentional content, is ascribed to it. So it should be *understood* as a *real* being.[153] Now of which forms of real being can we think? There may be many ways to divide this notion, but the way I propose here is chosen because it is very 'logical,' it is in accordance with tradition, and it serves the purpose of allowing for a systematic treatment of the distinction of the degrees of reflection.

A real being may either be considered as the result of *becoming* or not. If it is considered as the result of becoming it may either be thought of as itself *participating* in this becoming or not.

If something is the result of the becoming in which it participates, it cannot be completely determined in itself, for otherwise it could not *participate* in any becoming. But it cannot be completely undetermined either, for then it could not be the *result* of becoming. So it has a side of determinacy, which makes it a definite result, and it has a side of indeterminacy, which makes it capable of sharing in becoming. We call such a real being a *material* being, and of course we recognize in it Aristotle's unity of form and matter. The indeterminacy is then understood as the potency to change, the determinacy as the actuality of being this particular entity and no other. The determinacy makes it capable of being intended by the mind and located in a system of conceptuality. If this is to be done, abstraction must be made of the indeterminacy as a part of its reality, not of course of the *idea* of indeterminacy in general. Snow comes under the concept of snow as long as it is snow, and although the idea that it can melt *belongs* to the determinations of this concept, as soon as the snow has melted, the concept no longer applies to it. Because of this abstraction such a being is identified with its conceptual determination. It is immediately one with this determination. A rose is a rose is a rose, although by any other name it would smell as sweet. It is a qualitatively determined something. In ordinary life we consider everything to which we

---

153 In the case of mathematical objectivity this might seem problematic. Yet we do in fact consider mathematical objects as having an existence which is distinct from the existence of our act of conceiving them. That is precisely what we mean by their objectivity. If it were otherwise, the properties of mathematical objects would be arbitrary, which would make the whole enterprise of mathematics meaningless.

ascribe reality as a being of this kind. And what should be wrong with ordinary life? Why should this conception of reality be inferior to any other? No scientist can convince me that the chair on which I sit is more real as a complicated electromagnetic process than as the piece of furniture it is ordinarily taken to be. Not even Immanuel Kant can make me believe that it is more real as a product of synthesis under the category of substance than as a wooden chair on which I sit. Although the conceptualised world in which we live can, from a philosophical point of view, be considered as relative with respect to our faculties of perception and apperception, it remains as real in its relativity as it was before.

If something is the result of becoming in which it *does not* participate, it must be completely determined in its nature. It will be what it is, or it will not be. But as a result of becoming it has this latter possibility. Its becoming must be an *immediate* transition from not being into being. This transition cannot be a *process* as it is in material being. The existence and determination of such a being are indifferent with respect to each other. Therefore it is essentially *postulated*. It is not immediately one with its determination but rather *external* to it. But in so far as something is external to its own determination, it is external to itself; it is simultaneously completely determined and indifferent with respect to this determination. This kind of being is like the position of e.g. a planet in a coordinate system. The position is completely determined by its coordinates and yet the same position would be determined quite differently in another coordinate system. The determination is only what in German is called a 'Stellenwert' (a value of position) within a certain field of possibilities. Each possibility has the determinacy of a real being, but its reality can be either postulated or not. This kind of real being is an element of a structure, a quantitative being. It can be the result of measurement; and in the scientific world this is considered to constitute the highest degree of reality. But it is an immaterial reality existing within the material world. As mathematical objectivity it can be thought of as existing within an immaterial world of its own, but even this existence depends upon the use of material symbols, which, even if they are completely imaginary, still have material quality as their content. Although mathematical objectivity does not presuppose material being for its contents, it does for its existence.

If something is not the result of becoming, it exists as determinacy only. It may act as the determinacy *of* something existing in another way, but its reality does not depend on this. In so far as the human spirit is intrinsically independent of any material determination, it exists in this way. But the ultimate principles of material being and of mathematical structure must be also of this kind, even if we do not consider these forms of being as pure imaginations of the human mind. For e.g. determinacy and indeterminacy, if they are real, cannot be defined purely by the fact that they are relatively opposed to each other (for this relation must have a determinacy of its own) or from the way in which they are united in material or mathematical being (these ways being different). What they mean and what the nature of their relationship is must be independently known; and if it is really *known*, their meaning must necessarily be a kind of reality. One should neither confuse them in a Platonic way with mathematical determinations and imagine their existence in an ideal world of their own, nor in a subjectivistic way identify their reality with the existence of the knowledge of them in the human mind. Even less should one confuse them with material beings and regard them as 'qualia' or properties of something or even as a kind of substance (which is the common way in which the idea of their reality is criticized).

b) The epistemological side

On this side the relationship of perception and intellect is at stake. In ordinary life they form an immediate unity. We perceive that the grass is green and we are never aware of a distinction between the act of sense perception and the act of judgement involved in establishing such a fact. Yet the fact is not denoted by a mere cry, but by a sentence expressing it by means of words used as identifiers of general concepts. Epistemological investigation can teach us that establishing a fact is a very complicated happening, no matter whether you analyze it in the style of Aristotle or Kant. The ultimate effect, however, is that the individual instantaneous fact, which is also a stage in a process of change and therefore not completely determinate, is understood as an instance of a determinate general concept.[154] This is precisely the first degree of

---

[154] All major philosophers have expressed their view on this matter. Since discussing these views would lead to a volume of its own, I shall only indicate here in which respect I consider my view different from theirs. In Aristotelianism the existence of the *objects* of perception is not dealt with as very problematic. Kant, on the other hand, problematizes this existence in such a way that it is ultimately localized outside the phenomenal world. Hegel tries in his Phenomenology of Spirit to develop the epistemic relationship without falling prey to objectivistic or idealistic prejudices. Sense-experience, however, is considered too much in the service of understanding and reason. Modern phenomenology and philosophical anthropology, especially in Scheler and Plessner, counterbalance this intellectualistic one-sidedness, but lack a rigorous systematic coherence. Henri Bergson and his Dutch pupil H.M.J. Oldewelt occupy a special position in that they try to investigate pre-conceptual experience. This is what gave me the idea that the 'life-world' of phenomenology is itself already the result of reflection. In my opinion it is certainly true that philosophy, like art and religion, must necessarily make an appeal to pre-conceptual experience, but its investigation cannot be separated from reflection on the

reflection as described in Chapter 2. Its inner contradiction is that on the one hand reality is considered to be exactly as it is conceptualized, while on the other the conceptualization is known to be context-dependent. It serves the aims of a certain culture in a certain historical phase of development. It represents a 'power of tradition' precisely by creating a 'life-world' which appears as perfectly objective.

Mathematical thought in the modern form of experimental/nomological science is the logical outcome of the sublation of this contradiction. It creates a sharp opposition between the ideal objects of thought and the world of sense-experience. Measurement, at first in a literal, later in a wider sense, constitutes the connection between the two; and it is the very idea of measurement which contains the contradiction resulting from this attitude. For measurement is the interpretation of a real perception as an ideal object. In the perspective of the scientific attitude this is a category mistake. But without this category mistake science could have no results. Mathematical reflection understands the real principle of structurability in an ideal form. Its basic presupposition, therefore, is in contradiction to its attitude of separating the ideal from the real, the model from the phenomenon. It is the principle, which itself violates this separation, which is at the same time the foundation of its methodological fertility.

The sublation of the inner contradiction of mathematical thought, therefore, consists in the idea of a principle. Since this idea, as it has been explained in the previous chapters, includes the fundamental identity of experience and thought, it completes this particular dialectical development.

c) The existence of principles

The distinction of three meanings of the expression 'real being,' described above, includes an epistemological distinction, which has been developed dialectically, and an ontological distinction, which implies a gradation. Therefore I consider it as a possible solution of the problem to find a synthesis of the dialectical and the classical approach. There may be better or clearer answers, but the important thing is that attempts be made. It is difficult to avoid a very idealistic interpretation here, though I do not intend it. Remember that an analysis of the concept of 'real being' was the origin of my attempt. Of course the natural attitude of common sense thinking is to recognize exclusively material beings as real. In science, however, this attitude is already forsaken. Theoretical entities such as particles, wave-functions, transformation-groups, etc., which are essentially mathematical objects, are discussed as if they constituted the real substance of the material world, although they are dealt with mathematically. There may be a certain amount

---

conceptualized world. One should be on guard for a dualism of thinking and living.

of 'Promethean shame' among scientists, which makes them say comfortingly that these entities only exist within a *model*, but in fact they explain the material phenomena from the model, not the other way round. So *in practice* the model is considered to possess the higher form of reality. The world picture of modern science is essentially dualistic and idealistic. This idealism is in a certain sense justified. Structures are more real for our *thought* than perceptions, that is the source of all mathematism, and of all idealism in which the ideal is opposed to the real. Yet, precisely in so far scientific thought is also *empirical* thought, it has to admit that the *possibility* to explain phenomena from a model has to be tested. It is, therefore, believed that it *can* be tested, which means that structurability as a principle which unites the mathematical world of models with the material world of phenomena is fundamentally affirmed as the rock bottom of the possibility of scientific investigation. This principle cannot share in the relativity of either the material or the mathematical world. It must be absolute in the sense that its particular content does not depend on a certain context. What is considered as the reality of the material world depends on how we conceptualize our experience; what is considered as the reality of a mathematical world depends on what kind of structure we presuppose. But the origin of the *possibility* to conceptualize experience or to postulate structures cannot possess such relativity, for this would mean that there were contexts in which experience is *not* conceptualizable or *not* structurable. Moreover such a retreat to an even higher context is only begging the question.

But why should the principles of conceptualizability and structurability be the only absolute principles? It may look very Platonic to think of a 'world of absolute principles,' and in a certain sense it *is* Platonic, but everything depends on how we specify the mode of being of these absolute principles.

Absoluteness, in the sense of the absence of two *specific* modes of relativity, implies that principles cannot be confined to either the material or the ideal mathematical sphere. They are real in a sense which is beyond that opposition. Only if the 'ideal' is somehow unknowingly identified with the *mathematical* form of ideality can it be understood as *separate* from the real. The ideality of principles, however, is *identical* with their reality: it is their *activity*. The notion of activity can be used in a physical and in a mathematical sense, from which we have to distinguish its use with respect to principles. Physically the notion of activity is used in connection with a *process* in so far as it can influence other processes. Within mathematical thinking we do not use the notion of activity, but we could say that if, in some context, we think on the basis of a certain theory, the axioms of that theory are *active* in that context. We find both meanings of the word together in information-technology. We say that a hardware system is active if certain physical processes take place which support the use for which we have acquired the system. We say, however, that a computer program is active, if it is installed in such a way that the system works according to the rules determined by *structure* of this program. In this last sense we can also talk about the activity of a natural law: it is understood to determine the direction of a process *by its structure*. Every physical process can be regarded as determined by physical activities and at the same time by structures. In modern physics it is regarded as an axiom that these two sides perfectly correspond. In technology it is precisely this correspondence which is used to control natural processes. The correspondence is clearly regarded as a *reality*. But what kind of reality is it? Neither a physical activity nor a structure. It is presupposed as a real principle which determines the nature of physical processes in general. The principle is *active* in determining the nature of physical being in such a way that mathematical/experimental investigation is adequate. Processes of organic life may be determined by a somewhat different principle, which does not guarantee strict correspondence of action and structure. The structure of an organic whole in terms of its partial processes does not inform us sufficiently about the direction the whole process will take.[155] For this reason we talk about 'life' as if it is something which is active in organisms and from which we can understand their behaviour more adequately than from physico-chemical-cybernetical analysis, although by the latter method we can understand many things which are not immediately evident to us on the basis of the notion of a living organism alone. For in each being several principles can be active at the same time.

Principles, then, form the content of real being as it is intended in philosophical reflection. In this form of reflection the opposition of ideal and real existence is sublated. The intentional object is: intelligible real being. The existence of such an object is a presupposition of all ontology and all epistemology. It is unnecessary to remind the reader of the fact that also the denial of the possibility if either ontology or epistemology presupposes some ontology as well as some epistemology.

d) The congeniality of reflection and reality.

All knowledge is based upon intuition, that is, upon receptivity with respect to its object as real being. But human intuition has two handicaps.

---

155 Once I heard an zoologist who did experiments with rats sigh: 'Under precisely controlled circumstances the animal does precisely what it damn well pleases.'

1. It is *virtual*, which means that it becomes effective only by its theoretical and practical expression.
2. It is necessarily restricted by some perspective, which means that it can never encompass the total being of its object, but only grasp it in a certain respect.

In the following I shall summarize the principles constituting the three degrees of reflection. Of course these theoretical perspectives are by no means the only possible human perspectives. There are perspectives of conscience and value, of imagination and creation, of love and adoration, etc. But it remains true that everything that happens in life will also be expressed within the perspectives of theoretical reflection. Everything will be verbalized by the use of current concepts, investigated scientifically and philosophically. This does not mean that it is *reduced* or reducible to these perspectives, it only *appears* within them. In the same way the congenial object of each of these theoretical perspectives themselves can appear within the others, but with less adequacy. Mathematizing philosophy, for instance, is not impossible or completely nonsensical; it is only that its results can replace genuine philosophy no more than philosophical ethics can replace moral action or philosophical aesthetics can replace art.

The intuitive element of conceptual knowledge is realized in the meaningful conceptualization of experience within a certain culture. The possibility of different meaningful conceptualizations does not refute this view, as the existence of different possibilities to build a bridge does not refute the idea that it is useful to reach the other side of the river, and that not everything contributes to the reaching of this goal. The principles constituting the perspective of conceptualization are, as we have seen, the principles of *material being*, and the intuitive faculty congenial with them is usually called 'common sense'[156] which consists in the capability of conceptualizing a concrete experience *meaningfully* within its context.

---

156 This term is probably derived from the scholastic term *sensus communis*, which means the reflective component of sense-experience: by the sensus communis it is e.g. possible to experience a colour not only with respect to its quality, but also *as something which gives us a visual impression*. What is meant today by the term 'common sense' is, however, much closer to what the scholastic philosophers called *vis cogitativa*, the faculty of uniting sense impressions in such a way that they form an *object*. The activity ascribed to this faculty is usually compared to what Kant meant by *synthesis* and which by Kant as well as by the scholastics was ascribed to a cooperation of the senses and the intellect.

The intuitive element of mathematical knowledge is realized in finding structures which constitute an original starting point for reasoning about a problematic situation. One may think for instance of Einstein's theory of relativity or of Dirac's formalism for quantum mechanics, or, to give some older examples, of Euclid's geometry or Newton's mechanics. If different solutions are found to the same problem, the mathematical mind does not rest till they have been united into one theory. Here the leading principle is of course the principle (or principles[157]) of structurability, or of *material being as material being*[158] and the intuitive force is genuine mathematical intuition, the scope of which is by no means restricted to mathematics proper.

Philosophical intuition is a much-discussed subject. My view is that it is realized in the historical development of philosophical thought. In this respect it resembles conceptualizing intuition more than mathematical intuition, the latter being able to ignore previous attempts and to find completely new solutions to old problems. Philosophical intuition is also in a certain sense bound to a cultural context, but only with respect to *style*, not with respect to content. Its result is an anamnesis and coherent systematic representation of the leading principles implicitly enabling the cultural and historical situation in which it operates to exist. It is superfluous to say that I am referring to *real* principles, otherwise they could enable nothing. It follows from my line of reasoning that this too is a certain perspective, which must be constituted by its own principle. This, of course, cannot be anything less than the principle of principles,[159] or of *being as being*. What this means is itself the central question of a philosophical discipline: metaphysics. The formulation here of the principles of the three degrees of reflection now make clear what is the source of mathematism is: if *being* is identified with *material being*, the result

---

157 On unity and plurality of principles see Chapter 5.

158 See Chapter 1, especially section 4, in which the principle of structurability is understood as resulting from an idealization of the *material* character (determinability) of material being.

159 Husserl calls *intuition* the principle of all principles in his *Ideeen* §24. This is the epistemological counterpart of the Aristotelian ontological formulation used here. The term 'intuition' is used by Husserl in the sense of immediate knowledge of what itself is given. This indicates that core of all knowledge in which there is identity of knowing and being. If something is known intuitively, it is known 'as it is,' without abstractions or additions. Intuition, therefore it is the general capacity to grasp things as they are, and therefore it correlates with things as they are themselves: being as being. Of course this does not mean that *all* beings can be known in such a way or that we can be completely certain about the adequacy of the way we express what we take for intuitive knowledge. It only means that all knowledge must necessarily have an intuitive kernel.

will be that metaphysics is identified with mathematics for this identification implies the identification of being as being - the perspective of metaphysics - with material being as *material* being - the perspective of mathematical thought.

In Chapter 5 the disastrous consequences of this identification for metaphysics will be investigated. Such an investigation is not complete without an attempt to overcome the problem. This turns out to be extremely difficult because of our firmly established habit to grasp for structure as soon as anything becomes confusing. We want order in the situation and the first kind of order we reach out for is structural order, systematization. Real metaphysical questions, however, tend to escape such systematization. For the positivists this was a reason enough to regard them as meaningless, and in the perspective of mathematical thought they are indeed meaningless. The metaphysical perspective itself, however, cannot be regarded as meaningless, nor as unimportant, because its investigation questions the ultimate principles which govern our thought and action.

Chapter 5
Overcoming the mathematical paradigm in metaphysics

In the preceding chapters the distinction between empirical, mathematical and philosophical reflection has been developed. In this chapter the problem of metaphysical mathematism will be reconsidered on the basis of this development.

The paradigm of mathematical thought has in many ways influenced modern metaphysics as well as the criticism it evoked. The force and grandeur of mathematical thought is the explicitness and clearness of its particular presuppositions. Even if a mathematical theory has not been formalized, it is felt to be evident what is, and what is not permitted within its boundaries in order to try to solve its problems. This evidence seems to be the origin of the theoretical ideal behind all striving for 'exactness' in philosophy, and behind that kind of criticism on 'traditional' philosophy as well as phenomenological an hermeneutical philosophy that amounts to the reproach of 'softness.' On the other hand it is mathematical objectivity with its unchanging and purely structural nature, which allows for the foundational clearness of particular theories. This clearness, therefore, is not simply to be transferred to another field of knowledge, either empirical or metaphysical.

This accounts for another reproach often heard in connection with modern metaphysics, and, if its particular nature is not seen very clearly, even with all metaphysics. It is said that metaphysics is an expression of a striving for power. And indeed the mathematical method plays an important role in acquiring power over natural processes, therefore its metaphysical use can indeed been understood as a means for acquiring power in the spiritual realm. A mathematical proof is said to be compelling, and metaphysics, if it follows its paradigm, compels the mind into a certain direction. But in mathematics the free choice of a certain system underlies this compulsion, and in metaphysics we have to do with the absolute. Therefore it is feared that the power which is useful in the mathematical case, may be harmful in metaphysics. This fear however is based on the confusion of mathematical and philosophical reflection with respect to metaphysical questions. Although this confusion is systematically untenable, it has been historically meaningful in so far as it has made it necessary to make a clearer distinction between the degrees of reflection. In this chapter the nature and consequences of the confusion will be discussed and an attempt will be made to set metaphysics free from its mathematical inheritance.

5.1 Metaphysics in a mathematical style, and its fate

Mathematical abstraction results in structure, which is essentially one specific result of the possibility of many constructions, and therefore contingent. In Chapter 1 it is explained that mathematical structure is grasped by - ideal or real - actualization of a potency: structurability. This actualization essentially includes arbitrariness. It is true that specific systems of conceptuality resulting from the first degree of reflection can also be understood as constructions, but then they are understood from a mathematical point of view. In common sense itself, we do not experience concepts as the result of construction.

Philosophical reflection on the other hand aims at necessity, for the coherence of its objects - principles - cannot depend upon tradition, convention or postulation. Any blending of mathematical and philosophical reflection bears the suggestion that there exist necessary constructions, which is a contradictio in adjecto. So if metaphysics is implicitly contaminated with such a blending, it is an easy prey to criticism depicting it as either absurd or obscure. A construction has definite inner relationships, definite elements and definite properties. All these are definite, because they have been *defined* to be such as they are, and this means that there is arbitrariness in them. Principles nor their relationships, on the other hand, are the result of definition, they are on the contrary *presupposed* in any definition. They constitute the perspectives in which we can try to conceptualize or reconstruct experience. Their relationships are beyond definition, because they are constitutive for any definition. Nevertheless, in their implicit form, these relationships are better known than explicitly defined structures. They are implicitly but effectively known to us, and attempts to express them explicitly are experienced as highly artificial. They are not axioms, nor 'necessary truths,' nor expressible in a judgement or theorem without already presupposing them. We can investigate them, but never use them, apply them or draw conclusions from them outside the perspective they constitute. Yet, if we want to investigate principles, we must somehow express the results of this investigation. This is where the difficulties begin, for how to express such results in a form which must necessarily be determined either by tradition or by construction? Philosophy seems to hesitate continuously in its form of expression between mathematics and literature.

Literature is suggestive to us on the basis of culture and tradition. It can express truth, it can make one think, and it can point towards insights into necessary connections. But it lacks liability to critical investigation of its evidence. It either convinces or does not, but in the latter case one can rarely lay ones finger on the spot. Reducing philosophical prose to 'literary text' means depriving it of its real ambition: expressing intelligible necessity as such.

Mathematics, on the other hand, owes its intellectual force and its certainty to the immediate, conventional and systematic forms of representation of its objects.[160] In these forms of representation there exists a structural relationship between the intended mathematical objects and the way they are expressed. This specifically mathematical relationship of sign and meaning is not necessary in a strict sense, but it characterizes a mathematical discipline so strongly, that *within* the discipline it appears as necessary. Geometry without figures and algebra without formulas is not impossible, and in some periods of the development of these disciplines purely linguistic expression was even normal, but, as Leibniz observes, it is very hard in this way to travel a long distance without getting exhausted.[161]

Although Hegel explicitly rejects Spinoza's conception of philosophizing *more geometrico*, even in his own systematic way of expression we can find this mathematical element. He gives philosophical depth to the common sense divisions of a field of knowledge by developing it systematically, thereby using the agreement of form and content as a standard. He also uses a number of expressions, occurring all through his writings, like '*an sich*', '*für sich,*' '*external,*' etc, in order to indicate systematic connections between such developments in different parts of his system. It is very difficult to avoid the impression that this structure of expression has something to do with its content. In fact we have a relationship here, which is analogous to the relationship between representation and objectivity in mathematics. In mathematics this relationship is in a sense mathematical itself, whereas in Hegel's dialectics, this relationship is itself dialectical. It starts from the factuality of common sense conceptualization, then proceeds by making us feel the radical opposition between content and form of expression, but yet on another level we become conscious of the origin of this opposition, which is our tendency to ascribe an independent existence to the from of expression, thereby making it a content of our reflection, instead of understanding it *as* the expression of its original content. If he divides nature into the spheres of the mechanical, the physical and the organical, his foremost aim is to show us that in these spheres, distinguished by common sense, the *same* principle is at work. There is an opposition between the systematic diversity of moments, developed in the *Philosophy of Nature* and the radical unity of the intended principle: the Idea in the form of externality. Yet we know that diversity is the only way in which unity can be expressed, and as long as we do not try to separate expression and content, there is no contradiction in this. In the expression as expression there is a relative opposition between content and form, which does not exclude their identity. Although this motion of thought towards a philosophical level of expression is without doubt intended by Hegel, he does not always succeed in making the reader participate in it. The mathematical level is interfering and before we know it, we indulge in the beauty of the systematic development, and we fail to follow the direction

---

160   See chapter 1, section 2.

161   "Sans çela nostre esprit ne sçauroit faire un long chemin sans s'égarer" [Without that the mind could not go a long way without getting exhausted] G.W. Leibniz in a letter to Galloys from 1677. In: G.W. Leibniz, *Die philosophischen Schriften*, 1965.

of the finger pointing towards the real principle. All attempts to 'formalize' dialectics result from thought which has fallen victim to this temptation.

Philosophical systematization, however, cannot aim at representing certain structures in such a rigorous way. It has to transcend its own particular structure, not into a literary expressive imagination, but into the intellectual challenge of its proper aim: establishing real insight. Such systematization has the function to prevent thought from stopping at too low a level of understanding, it provides the formulations of the problems, but it is never itself a solution. Philosophy is the encouragement of the intellect to recognize that it knows more than it thinks it knows. Participating in philosophical thought always requires that we give up some prejudice concerning what we imagine to be the definition of 'knowing.' This distinguishes the intellectual challenge in philosophy from its mathematical counterpart. In mathematics the challenge is directed towards the faculty of imagining of and reasoning about new and unheard-of structures. The principle of 'knowing' in mathematical reflection, however, always remains the same: structurability. In philosophy there is no fixed principle of knowing, only the attempt at explicitly knowing the principles guiding all of our knowledge. The 'exactness' of philosophical expression, therefore, is of a negative nature. Its function is to prevent a premature feeling of understanding. All beginning students of philosophy complain about this. They justly feel that philosophical language aims at making things more difficult instead of easier. Why cannot this be said in a more simple way? In a certain sense this resembles the situation in mathematics. There too things are said in a complicated form. But one feels that the reason for it is, that they *are* really complicated. In philosophy, however, anyone who has feeling for what it is all about, becomes convinced that understanding the complicated cannot be the ultimate aim here. Principles must be simple, and it is because of their simplicity that it is difficult to grasp them. So why cannot simple things be expressed in a simple way? The answer of course is, that simple expressions suggest to the untrained the wrong kind of simplicity of the content. In 'occult disciplines' of certain religious societies this is no problem, because the expressions are only meant for the initiated, who are supposed to understand them properly. Philosophy, being a *rational* discipline, must necessarily provide its own initiation. It cannot separate a cult of initiation from the expression of its actual contents. In a rational discipline one becomes initiated because one takes up the intellectual challenge, one understands what is interesting about it, whereas in initiation rites one participates not in the first place because one understands what they are good for, but because someone with authority says they must be undergone in order to understand what they were good for afterwards.

Mathematical and philosophical *expression* have, as we now understand, diametrically opposed criteria of adequacy. Mathematical expression is better, in the measure in which it allows us to connect mathematically the structure of our language with the structures expressed in that language. The more rigorous this connection becomes, the more our way of expression gains the character of a formalism useful for accurate proof and computation.[162] Philosophical expression, on the other hand, is better in the measure in which it prevents the intellect from clinging to certain definite structures of knowledge and self-expression. Mathematical language should enable us to concentrate on definite structures, philosophical language should prevent such concentration with the aim of opening up our minds for the origin of our perspectives without presupposing any initiation into extraordinary realms of experience.

For this reason any attempt to develop metaphysics following the mathematical paradigm must necessarily end in the fundamental rejection of all forms of metaphysics. The dilemma between the mathematical and the philosophical criterion of adequacy of expression is unsolvable. There is no dialectical solution either, because to choose for dialectics is already to choose for the philosophical criterion. If dialectics is understood as a formalism, complete rejection of it is not far behind.

On the one hand to choose the mathematical criterion for philosophy, leads to nihilism. If, on the other hand, mathematical reflection acquires metaphysical pretensions, it cannot very long remain content to be pure mathematics. It has to incorporate some philosophical reflection, and in the measure in which it succeeds in expressing this incorporation explicitly, it disqualifies itself as mathematics. In the measure, however, in which it succeeds in satisfying the requirements of mathematical expression, it becomes philosophically irrelevant. In its naive form it becomes dogmatic because it postulates some more or less arbitrary constraining framework, which nevertheless is infected by the metaphysical claim that it expresses necessity. In reaction to this dogmatism it then becomes nihilistic, for the arbitrary character of the construction is brought to the foreground. It will be insisted then that 'anything goes.' In this case, 'anything' is not really anything of course, but

---

162   Compare the development of forms of mathematical representation in chapter 1, section 2. The immediate, conventional and systematic representation are in this order increasingly adequate as mathematical forms of *expression*. This does not mean, however, that the systematic form is always superior as a form of notation for the working mathematician. The immediate form is often more inspiring for mathematical intuition, but once the problem has been solved, the mathematician tries to find a rigorous proof is a form which is as systematic as possible.

any *construction*,[163] and that is not what we are looking for in metaphysics. Therefore this trail leads us into nothingness. Not only by its own logic we can understand that the mathematical paradigm in metaphysics must lead towards nihilism, it can also be seen to have happened in fact.

5.2 The influence of the mathematical paradigm on metaphysics

In antiquity the mathematical paradigm had a certain influence on philosophical thought, but this influence was generally consciously affirmed. The Pythagoreans really felt that number was the essence of everything and Academians such as Speusippus and Xenocrates consciously interpreted Platonism in a mathematical way. In modern times, however, the influence is not consciously recognized. Descartes was fully aware of the distinction between metaphysical and mathematical questions. Yet he determined the objective world as *res extensa*, because his line of thought led in that direction.[164] The influence of the mathematical paradigm had already been decisive in a far earlier stage of his thought. His decision to find an absolute starting point from which his whole philosophy could be developed was thoroughly influenced by it. But this decision also fits very well in the whole atmosphere of his method: 'do not rely on authorities, think for yourself.' Even our own intuition should be distrusted as an authority as long as it is not thoroughly tested as we do in geometry by giving proofs or making calculations. Fundamental ideas should not only be 'clear,' but also 'distinct.' The kind of certainty Descartes is looking for is mathematical in nature: a proof by distinct calculation from a clear starting point, giving a clear result. His analytic geometry is paradigmatic in this respect. Geometric insight and calculation have to complement each other.

But empiricism is not free from the influence of mathematics as a paradigm either. The experiential basis of knowledge it is looking for, must have the same characteristic as Descartes' clear and distinct ideas. It should be a rockbottom basis for all reasoning, and not, like real experience, liable to an infinity of interpretations. It consists of simple elementary sensations or ideas, from which more complex notions are constructed by combination. All possible interpretations of the elementary data have to be thought of as of constructions and combinations. Therefore all intentional content of perceptions has to be understood as resulting from constructive interpretation. What in fact

---

163 Cf. B. Taureck, *Das Schicksal der philosophischen Konstruktion*, 1975.

164 Compare chapter 3 section 3.

remains of external perception is the inner perception of its psychic experience. Locke performed this reduction only for the so called secondary qualities, but Berkeley already saw that Locke's argumentation held as well for the primary qualities.[165] Only the *form* in which we experience what must qua content be regarded as complex, is now left as the rock-bottom basis of our knowledge. Any content of experience depends on our conceptual apparatus and must therefore be considered as 'impure experience.' A colour cannot be given a definite name, because that name is based on a division of colours. What remains is in this case colour as such, which is the *form* of the experience of the particular colours. What has to be the experiential basis of all knowledge in empiricism can now be seen to be an indefinite plurality of infinitesimal sensations or ideas, which is as abstract and formal as Descartes' *res extensa*. The real externality of the object of the natural sciences is reduced to a mental construction on the basis of an 'inner externality' within the psyche of the observer.

Modern philosophy can be seen as the attempt to express in a philosophical way the presuppositions of modern natural science. This can be recognized in two distinctions returning in various forms in modern thought: The distinction already mentioned of primary and secondary qualities, which essentially corresponds to my distinction between structurability and conceptualizability. The other distinction is expressed by Hume as the

---

165 In modern philosophy the 'primary qualities' are thought of as precisely the quantitative (structural) properties of material being. The 'secondary qualities' are the perceived properties in so far as they are not similar to their real physical causes, which are supposed to be the primary qualities of a purely physical world, which according to Locke had to be independent of our mental acts. Berkeley rejected the hypothesis of this independence and regarded the physical world itself as an 'idea.' Because in empiricism the secondary qualities are understood as non-intentional psychical affections, it cannot avoid psychologism in its epistemology. This element of empiricism is thoroughly criticized in Edmund Husserl's *Logische Untersuchungen*, 2. Band, Teil II. Although the idea of a purely extended physical world, in which matter has no other than geometrical qualities (except for the implicitly assumed mutual exclusiveness of place) goes back to the ancient atomists, the use of the *terminology* of primary and secondary qualities in this context is usually ascribed to Boyle. The terms in fact come from the Aristotelian tradition, in which Galenus introduced them, but in which they have a quite different meaning. In this tradition primary qualities are those qualities which are perceivable by tactile sensation only. They are called primary, because all senses are supposed to be transformations of tactility. Therefore the qualities perceivable by these other senses are considered to be modifications of the primary qualities. It is imaginable that the modern preoccupation with mathematics and mechanics has led to the identification of the tactile with the extended, which seems to be implicitly presupposed in the transition to the modern use of the terminology of primary and secondary qualities.

distinction between 'matters of fact' and 'relations of ideas,' and by Leibniz as the distinction between '*vérités de fait*' and '*vérités de raison*.' If one looks at the examples of 'relations of ideas,' and of '*vérités de raison*,' they are mainly mathematical. This distinction too seems to express rather the distinction between the first and second degree of reflection than anything else. And in fact it really is this distinction which is of paramount importance to modern natural science. *Making* this distinction *explicitly* is an activity of philosophical reflection. But this reflection seems to lack a territory of its own in the modern era.

In order to avoid an extensive study of modern philosophy in this context, which would fill a whole book by itself, and which has also been partially done by others[166], I shall now jump towards a philosopher who, least of all the moderns, could be suspected of mathematism: Hegel, who explicitly rejected the mathematical method in philosophy although he had very great respect for Spinoza.

Hegel is still influenced by the mathematical paradigm of the moderns, and this influence seems to be present strongest precisely where he tries to avoid it. It is understandable in this connection that Hegel has much difficulty in determining the place of mathematics in his system, because he cannot see it otherwise than as an abstract - and therefore inadequate - form of either philosophy or physics.[167] This aspect has already been dealt with in chapter 3. Here we shall take a closer look at the rigour of his dialectics and approach the subject of Hegel's hidden mathematism from that angle. It is the magnificent coherence of his systematic thought which has bred its fiercest enemies. A serious student of Hegel's works is necessarily impressed by the amount of truth contained in them. Most criticism following the method of lifting out a passage and ridiculing it is based on historical ignorance concerning the context in which such passages have to be interpreted[168]. Why does this monumental philosophical oeuvre mark the *end* and not the *beginning* of a philosophical era, or in other words, why was it rejected towards the end

---

166   See for instance: Taureck B., *Mathematische und tranzendentalen Identität*, 1973.

167   See also my articles on Hegel's attitude toward mathematics and mathematical physics in *Hegel und die Naturwissenschaften* and *Hegel and Newtonianism* both edited by Michael Petry

168   Concerning the Philosophy of Nature, recent investigations have confirmed this. See e.g. Horstmann R.P. and Petry M.J., *Hegels Philosophie der Natur*, 1986, and Petry M.J., *Hegel und die Naturwissenschaften*, 1987.

of the nineteenth century as a basis for further philosophical development? One of the reasons may have been that is was felt to be too much of a good thing. Although Hegel certainly meant to express the motion of living philosophical thought, the result seems to be infected by the spirit of heaviness. The structure of the dialectical development becomes a burden to philosophical thought because of its seemingly mathematical rigour. Even nowadays excellent philosophers try to decipher its formula[169]. But could it not be the case that in spite of Hegel's explicit rejection of Spinoza's method of philosophizing *more geometrico*[170] the mathematical paradigm has found its way via Kant from modern rationalism and empiricism into Hegel's thought? The central phenomenon from which reflection in this philosophical tradition takes its origin is the possibility of modern mathematical natural science. This implies that what has to be explained is the possibility of knowledge of what is understood as *external* to human intelligence. Kant's *Stoff der Erfahrung* (matter of experience), the totality of sense impressions, is understood as external to the transcendental unity of apperception. In the Critique of Pure Reason this externality is taken as a fact, but Hegel wants to understand it. For him ultimate reality must be the unity of the internal and the external. His Science of Logic aims at developing philosophical insight into such a notion of reality, or rather: of being as being. Did he succeed? The formulation 'absolute Idea' expresses his claim that he did, but the dilemma is whether we should understand this as an anthropological or as a metaphysical claim. Does Hegel's system only describe the way in which the human mind necessarily has to express itself in an external context in order to develop historically towards autonomy, or does it indeed claim to understand all reality from an insight into the coherence of all metaphysical principles, 'God's thoughts before the creation of the world' or the absolute idea as it is described in the Science of Logic?

The suggestion is very strong that the absolute idea, binding together all principles like a 'one ring[171]' is not only completely intelligible to us, but also justifies the specific structure of Hegel's system. Yet, as it is already remarked in chapter 3, this suggestion is somewhat misleading, because Hegel himself did never hesitate to make additions and correction in his system. The problem is, that in its systematic structure, there is no place for expressing this openness with respect to the structure itself. This is a curious paradox: the

---

169   See e.g. Vittorio Hösle, *Hegels System. Der Idealismus der Subjektivität und das Problem der Intersubjektivität*, 1987.

170   In the preface to the *Phenomenology of Spirit*.

171   See J.R.R. Tolkien, *The Fellowship of the Ring*.

system aims at the expression of the transcendental openness of the human mind, that is its ability to grasp transcendental principles by intellectual perception; yet it is not able to express the openness of the philosophical method by which it is composed! The system therefore acquires traits of a 'necessary construction,' which is precisely the impossible contamination of mathematics and metaphysics which tends to discredit all modern metaphysical positions.[172] This seems to be the basis of the widespread present consensus on the impossibility of such systems.

But the *'faute hypercorrecte'* is as usual in philosophy as it is in practice. Because it has not become clear that it is the mathematical paradigm which constituted the trap of Hegel's system, philosophical positions opposing to German idealism, such as Marxism, vitalism, existentialism, positivism and structuralism, however critical they are with respect to the modern tradition, are by no means free from this same paradigm. Essentially they all switch from the dogmatic form, which suggests a necessary construction to the voluntaristic form, which essentially expresses an abstract notion of freedom, such as only the mathematician possesses in relation to the sphere of ideal structures. Those philosophies all present themselves as absolutely valid insights, but reject any claim to knowledge of the absolute. To such philosophical currents, metaphysics counts as an ideological claim to authority which hampers human freedom. The 'necessary construction' is deconstructed and shown to be only one of infinitely many possible ones. Dogmatic systematics has passed into dogmatic nihilism and the project of metaphysics as such has become suspect.[173]

---

172 Heidegger's notion of *'Seinsvergessenheit'* can be interpreted as a philosophical expression of this confusion. It is coined to criticise ontological fixation of the opposition of subject and object. Such a fixation is a characteristic of mathematical reflection. It seems to be this mathematical element in modern metaphysics, which falls under Heidegger's criticism, and that makes it also clear why he understands modern technology as the realization of such metaphysics. The *'Verdinglichung des Seins,'* the blurring of the ontological difference, reminds us of what is done in mathematical reflection: creating ideal entities as actualizations of a potency. This potency - structurability - is of another order than its ideal actualizations - the mathematical objects -, and it is indeed 'forgotten' and inexpressible in mathematical thought. In the mathematical degree of reflection mathematical being as such is indeed in a certain sense *absent*, but absence cannot be written on the account of ancient and scholastic metaphysics: as a metaphysical trend it is thoroughly modern. Heidegger understood rightly that the confusion of the mathematical and the philosophical degree of reflection, which he did not interpret as a confusion but as a fate - *Seinsgeschick* -, must necessarily lead to nihilism.

173 Th. Adorno expressed the suspicion that this anti-metaphysical trend has been a process of flight from something which could not be left behind. "The process by which metaphysics continuously ended up where it was conceived to lead away from, has reached

5.3 Against metaphysics

That the anti-metaphysical feelings, common in our century, are really only an effect of the above-mentioned comedy of errors, can also be seen by a brief analysis of the arguments in which those feelings are rationalized, and which are all based on caricatures of the metaphysical project.
The arguments for rejection of 'metaphysics' can be divided into 7 kinds:

1. Identification of metaphysics as such with a (conception of a) particular historical metaphysical doctrine or cluster of doctrines, which leads to a rejection of metaphysics on the basis of arguments against these specific forms of it. I shall name such arguments *historical*.

2. Ascribing to metaphysics certain claims, which, on epistemological grounds, are to be considered as beyond our reach. I shall use the shorthand name *epistemological* arguments in this case. A special form of the epistemological argument is the historicist argument, which ascribes to metaphysics the claim of transhistorical and transcultural validity of its explicit expressions.

3. The identification of the most fundamental with the trivial, which does not need any investigation, but is clear to everybody. I call this the *futility* argument. It can also be accompanied by the idea that such knowledge is superfluous, because it does not help us to solve any practical, or even theoretical problems, neither in science, nor in technology or practical life.

4. Rejection of the *project* of metaphysics on the ground that an intellectual investigation of the domain of the most fundamental reduces this domain to something which is intelligible for the human mind, thereby doing injustice to it. This argument differs from the epistemological argument, because it does not imply that one *cannot* understand the most fundamental, but that one ought not try to do so. I call this the *ethical* argument. If other means of access to the domain of the most fundamental are recommended, the ethical argument can take the form of a *theological* or a *mystical* argument. There is also a *political* variant, suggesting that metaphysics with its emphasis on principality leads towards extreme positions and fanaticism.

5. The argument that metaphysical questions cannot be meaningfully formulated in any language. All existing formulations can be shown to be

---

its vanishing point" [Th.W. Adorno, *Negative Dialektik*, 1966 p.356.]

meaningless, and the whole domain of the most fundamental is only appearance, therefore it cannot be investigated. This will be called the *linguistic* argument.

6. The criticism of the tendency of metaphysical results to block further research, for whatever we have to say about the most fundamental tends to inherit some claim of absoluteness from its subject-matter. Knowledge of the absolute tends to present itself as absolute knowledge. I call this the *fallibilistic* argument.

7. Reduction of metaphysics to its possible social, cultural or psychological function, whereby it is judged independently of its epistemic claims. This is called the *functionalistic* argument. In this same line criticisms of metaphysics as being 'ideology' fits in.

In practice these argument are almost never found in a pure form. Voltaire's brilliant combination of the epistemological and the triviality argument is well known: "Metaphysics contains two parts. In one part is said what everybody already knows, in the other what nobody can ever know." Such combinations seem to strengthen the argumentation because of the complementary impact of the different arguments. It is a decomposition-technique of argumentation in which the possibility of the decomposition is usually overlooked as a presupposition.

In this context all the types of arguments will be dealt with separately.

ad.1 The historical type of argument, cannot reasonably content itself with the general remark that metaphysics is *passé*. It has to justify itself by proving the impossibility of specific historical metaphysical doctrines. Success in such an attempt does not prove the impossibility of metaphysics as such, but is on the contrary itself a contribution to it. The historical argument is fruitful for metaphysics. An example[174] is Heideggers criticism of the onto-theological conception of being, which has inspired neo-scholastic philosophers to re-investigate scholasticism in order to bring into recollection the original meaning of the medieval metaphysics of being, which was not at all what Heidegger made out of it. A similar process is stimulated by the post-modern criticism of the idea of the unity of being, ascribed to ancient *and* modern metaphysics,

---

174 An older example is of course modern metaphysics as a reaction to the rejection of Aristotelianism in the Renaissance.

and indiscriminately to all metaphysics. This has inspired new investigations into the meaning of the traditional problem of unity and plurality.[175]

ad.2 The epistemological argumentation against metaphysics can take various shapes according to the kind of claims ascribed to metaphysical knowledge. The most radical form of this type of argument is based on an (implicit) definition of metaphysics as being 'knowledge of that which is unknowable to us.' It is clear that by the contradictory nature of this definition metaphysics is to be considered as an absurd enterprise. Now this argumentation seems to be rather obtuse, but it lurks in the background of many discussion within the tradition of modern philosophy. Since Descartes questioned the possibility of immediate knowledge of the world as it is in itself, modern thought is haunted by the idea that something might exist, which is in principle beyond the reach of our knowledge. The notion of metaphysics therefore tends to acquire a double meaning. This can be demonstrated by many texts, of which this remark of d'Alembert's is only an example:

> Metaphysics is, according to the perspective from which it is considered, the most satisfactory or the most void of all kinds of human knowledge. The most satisfactory kind, as long as it does not consider objects beyond its horizon, as long as it analyses its subject matter with acuteness and clearness, and so long as it does not transcend that which it recognizes clearly in it. The most void it can be concluded to be, if, simultaneously audacious and obscure, it enters an area which is hidden from its view, if it disputes the attributes of God, the nature of the soul and of free will, and other themes of this kind, in which all of the philosophical past has lost its way, and in which modern philosophy cannot hope to have any better luck.[176]

As in the second sense, completely apart from what is supposed to fall under its domain, metaphysics is to be considered as being impossible by definition, in the following only the epistemological claims implied by the first sense will be dealt with.

Even if metaphysical knowledge is understood as insight into the real principles forming a basis of the perspectives from which we can investigate

---

175 See e.g. Karen Gloy, *Einheit und Mannigfaltigkeit*, 1981 and Jan Aertsen, Denken van de Eenheid, 1990 pp. 399-420

176 J. d' Alembert, *Oeuvres complètes de d' Alembert*, I, 1967, par. XV.

the world, the objection can be made that such fundamental insight is not within our power. All we know, it is said, is relative with respect to certain *methodological* principles, which we can only ascertain by our feeling of success in theory or practice. We can never be sure that these methodological principles are based on real principles, and that our feeling of success is real success.

Now this can easily be admitted by a proponent of metaphysics. It does not imply, however, that *investigation* of what is silently *presupposed* as being real principles and real success is superfluous or impossible. We cannot know more than we really do, but we always presuppose more than we really admit. And maybe we can even understand that these presuppositions have their own inner coherence or even necessity, making them worthy to be presuppositions. Such an insight is not at all excluded by the methodological relativism which is the basis of this objection.

But if there are really no presuppositions? If our methods are purely the result of a historical process of development? Then the objection turns into a historistic one:

Knowledge of the absolute is impossible, because all knowledge is determined by the historical situation in which it is acquired. So knowledge of a content which is, by definition *not* determined by a historical situation is excluded by contradiction. This argument hinges upon the meaning of the word 'determined.' It can be applied to the *material object* of knowledge: that of which something is known; to the *formal object*, that which is known *of* it; and to the form of *expression*, that what is said about what is known of something. Metaphysics seems to be based on a denial of the first sense of determination, whereas it would not be a meaningful enterprise if there were no historical determination in the second and the third sense. If it could not increase knowledge of its object or could not strive for better expression, there would be no possible progress. If however, it is claimed that the absolute *itself* is historical, this is to be regarded as a very interesting metaphysical position, which, however seems very difficult to defend.

Then, of course there are forms of the epistemological argument which are based on a particular epistemology. We need not say much about them here, for they either are dogmatic or presuppose some metaphysical foundation.

A more refined form of the epistemological argument consists of an implicit or explicit denial of the *existence* of the object of metaphysics. There is no absolute, there are no real principles, no essences etc. It is an obvious fact that such a denial itself implies a metaphysical doctrine. In order to avoid this paradox, it has been argued that we cannot in principle *know* whether absoluteness or necessity exists. But what is the status of our knowledge about what we can or cannot *in principle* know? The paradox is still there.

Ad.3 The futility argument is hard to counter, for in fact it is no argument at all. It is only a feeling. For how can one know that metaphysical investigation has never contributed or will never contribute to useful insight? Historically one can argue both ways. Greek philosophy may be construed to be the main source or to be only an epiphenomenon of the development of western science and culture. Science may be proclaimed to be completely independent of, or to be deeply dependent on metaphysical ideas. There are many convictions in this domain, but no single proof. My conviction is, that human knowledge is more coherent than one usually supposes. If one investigates a historical period, for instance the Renaissance, it becomes increasingly clear that everything interacts with everything. Science, philosophy, economy, technology, religion etcetera are not completely separate domains. Their development is so strongly interrelated, that it is useless to proclaim the development in one domain as the cause of all other developments. But it is as useless to proclaim certain domains - such as metaphysics - to be of no consequence. It is, on the contrary, very plausible that investigation into the most fundamental is very influential in all other domains. For precisely that which is regarded as evident or even trivial *after it has been discovered*, is most frequently adopted by others.

The futility argument, by the way, has some hypocrisy in it, for why object to a discipline because of the evidence of its results? Or is there some fear that one's own evident presuppositions will be scrutinized?

Ad.4 The ethical, mystical or theological argument has been dealt with extensively by scholastic philosophers, because philosophy was widely regarded in their time as a pagan business, originating in ancient Greece, and not in accordance with Christian dogmatism. A proper domain for the 'natural light' of the intellect had to be conquered, which was limited by revelation and by mystic knowledge. This resulted in the conception, foreboding the modern era, that the intellect had its own, natural restrictions, which need not and should not be enforced by external prohibitions. In the transition from the seventeenth to the eighteenth century the consequences of enlightment evoked mystical counter-movements within several religions (Cabbalism, Chassidism, Pietism, Jansenism etc.), introducing again the idea of a limitation of the freedom of thought. On the other hand a monstrous alliance between mysticism and positivism already appeared in the sixteenth century. Francis Bacon's prophecy: "Do not care about metaphysics. There will be none left when true physics has been found, beyond which will only be theology." is the best known formulation of it. In our century this attitude appears itself in philosophical form. It tends on the one hand to reduce science to a mere stocktaking of facts, on the other to reject all real philosophical reflection, but not mystical experience. Thus a new dualism is introduced, of which Wittgenstein's

Tractatus Logico Philosophicus is as clear a philosophical expression as Descartes' *Discours de la Méthode* was of the old dualism.[177] This attitude, however, did not result in real modesty, but on the contrary in a form of hypocrisy, motivating scientists and philosophers to banish the expression of their more speculative thoughts to their memoirs, thereby refraining from developing them systematically. It becomes clear from this historical sketch, that this kind of argument is rather an attitude, based on fear for falseness. I do not see any other way to counter this fear than facing it and investigating all possible points of view.

Fear also is at the root of the political variant. It is overlooked thereby, that it is not the philosopher who is in danger of becoming a fanatic, but much more the man of practice, who is always tempted to monopolize one perspective in order to ascribe the other ones to his adversaries and fight them.[178]

Ad.5 The linguistic argument presupposes a theory about the nature of language. If such a theory is more than a simple paradigm, it involves certain philosophical perspectives, the principles of which ask for metaphysical investigation. There is a 'pragmatic' contradiction of form and content if the theory really proves that such investigation is impossible, for thereby it endows its own principle with a dogmatic status. If there is only a paradigm of the nature of language, usually either the language of 'common sense,' or the language of mathematical reflection serve as the origin of this paradigm. Indeed there is a real problem of finding a proper form of expression for philosophical reflection. This problem would be unsolvable if language were not very flexible in its use. This is clear from the possibility of poetry, which does not mean that philosophical prose is a kind of poetry, but only that language expresses more than the contents of simple judgements on facts or illocutionary acts. Scholastic philosophers already understood that even such simple judgements had a deeper dimension. Judgement deals with the very being of something.[179] The conceptualization of experience presupposes for its meaningful use the reality of the implicit perspectives from which we conceptualize. Whorf is right in this respect. There is an implicit metaphysical

---

177   Since Ray Monk's excellent biography, Wittgenstein's own religious motives have become better known. Ray Monk, *Ludwig Wittgenstein, The Duty of Genius*, 1990/1991

178   Compare Hegel's 'column': *Wer denkt abstrakt*, Werke 1970, Band 2, pp. 575-581.

179   *Prima operatio respicit quidditatem rei, secunda respicit ipsum esse*, Thomas v. Aquino, I Sent. d.19, q.5,a.1,ad 7: The first operation (of the mind: conceptualization) deals with the nature of something, the second (judgement) with its proper existence.

dimension in language. Only he should have taken the idea of metaphysics more seriously. We can use this metaphysical dimension reflectively as a means of philosophical expression. This does not imply of course that the only thing we can do is reproducing the implicit metaphysics of common sense. In the realm of poetry this would mean that a poem could only express trivial thoughts and in the realm of mathematics only the conceptual structures of quotidian language could be reproduced. We do not deal with the depth-dimension of language in a purely reproductive manner, we deal creatively with it.

A special form of the linguistic argument is the therapeutical one. "The metaphysician is treated no longer as a criminal but as a patient: " .....there may be good reasons why he says the strange things he does." is A.J. Ayer's characterization of it.[180] This is a well known debating trick: to play the man instead of the ball. In the context of the two and a half millennia of history of this mental disease, I invoke the right to suffer from it without being bothered by therapy.[181]

Ad.6. The fallibilistic objection against metaphysics is based on the idea that metaphysical knowledge would always claim to be unfalsifiable, and therefore hampers further research.

It cannot be denied that metaphysics is in search of knowledge of the absolute and the unconditionally necessary. But knowledge of the absolute need not be absolute knowledge. In what we say and think about the absolute and the necessary, we can be as dead wrong as in what we say and think about what is relative and contingent. The claims of metaphysics do not concern the certainty of its knowledge, but the nature of its object.

But even if this is admitted, it can be objected that metaphysical knowledge at least claims to be unfalsifiable by *experience*. In a certain sense this is true. Metaphysics is in search of insights which are *a priori* precisely in this sense. Insight into principles constitutes perspectives enabling us to interpret experience. These perspectives are considered to be universal, in the

---

180   A.J. Ayer, *Introduction to Logical Positivism*, 1959, p. 8

181   This does not imply of course that there is no such thing as '*Die Verhexung unseres Verstandes durch die Mittel unserer Sprache*' [The spell put on our understanding by the means of our language] (L. Wittgenstein, *Philosophische Untersuchungen*, 109). According to Wittgenstein philosophy is the struggle *against* this 'Verhexung,' or as he writes elsewhere 'against the philosopher in ourselves.' Such passages express the dynamic character of philosophical thought. It has to express itself, but thereby creates the danger of becoming entangled in its own way of expression, which it will always be obliged to transcend. As I understand it, this was not something from which Wittgenstein wanted to be cured.

sense that any content of experience is in principle interpretable in it. But it is *not* claimed that the perspectives are universally *adequate*. If some perspective happened to be so, we should not need any other.[182] The notion of a perspective implies restricted adequacy with respect to experience and it is always possible to find new perspectives which give a more adequate interpretation of some experience. The *a priori* nature of principles only implies that a perspective cannot *lose* its adequacy either by experience or by reflection. So the unfalsifiability does not hamper research, as it is suggested in the objection, for one can always look for more adequate perspectives with respect to certain phenomena. What has been clarified, can always be clarified further, a fact which is overlooked by those who hold that only the refutation of all past philosophy can open new ways.

Ad.7 The functionalistic argument has the same catch as the kind of linguistic argument that claims to be based on a theory of language. If metaphysics is said to fulfil a certain function in society or culture, this idea must be based on a sociological or historical theory. The perspectives in which this theory explains or describes the world must be based on certain principles, open to metaphysical investigation. If it is argued that such investigation is of no consequence because of the function ascribed to it by the theory, this means that the theory ascribes dogmatical status to its own principles, and it is therefore based on a rather primitive form of metaphysics itself. This does not imply that it cannot be true as a sociological or historical theory. But this is irrelevant for the appreciation of the functionalistic argument. It may well be that metaphysics fulfils certain functions in society or history, but so do sociological and historical theories. This cannot, however, affect the value of their results in the perspective of knowledge, otherwise they would run the risk to refute themselves by their own success.

The rejection of metaphysics as 'ideology,' if it is not purely based on prejudice, must be based on some theory of society, such as historical materialism. Sir Karl Popper's heroic attempt to criticise such theories as 'closed' and 'totalitarian' hits the mark here in so far as it is true that such theories try to make themselves immune to criticism. But Popper is, in my opinion, rather naive as to the criterion that enables us to distinguish between genuine philosophy and totalitarian social or psychological theory. Neither of these is falsifiable by mere facts, but only the latter must defy metaphysical investigation of its fundamental notions. The ideology here is not metaphysics,

---

182   Of course there are universal perspectives such as the perspective of logical reasoning, constituted by the principle of non-contradiction. But they are not universally adequate in the sense that by such reasoning all questions can be answered.

but its rejection. Metaphysics itself, just like empirical science, is a potentially critical enterprise with respect to ideologies.

It is clear from this brief discussion that the debate pro or contra 'metaphysics' is not so much a question of rational argumentation as a question of attitude and feeling.[183] Yet it has a remarkable historical dimension. Antique scepticism and Pyrrhonism were motivated by an ethical aim: happiness of the individual, and they did not only reject metaphysics, but all knowledge. In the modern era, however, the rejection of metaphysics is strongly linked to the rise of the mathematical sciences. It is the method of these sciences, explicit mathematical reflection, that has to distinguish itself from and defend itself against the claims of the other degrees of reflection. Common sense, of course cannot be bluntly rejected, for all of mankind lives by it. Therefore metaphysics - including pre modern forms of physicalism, except mechanistic forms, which were closer to the mathematical approach - has been depicted as the enemy of the mathematical method, and the surreptitious interpretation of it from a mathematical perspective as a 'necessary construction' became the main weapon against it. This is the real argument,

---

183 This aversion to metaphysics is characteristic of the whole modern era. That it is not the privilege of the 20th century can be learnt from quotations such as the following: "...of instances, formalities, quiddities, haecceities, in short, of things no man ever can look upon, unless he be a Lynkeus, being able to see nothingness through sheer darkness." (Erasmus, Welzig 2 (1975) p. 133); "Peripatetici, ne rigentono solo il nome, contenti, senza passegio, di adorar l'ombre, non filosofando con l'avvertenza propria, ma solo la memoria di quattro principii mal intesi" [Peripathetics, who do not even deserve this name, because they do not go about but are content to adore the shade, do not philosophize with due care but only by bringing into remembrance four principles, which they understood wrongly.] (Galileo, Opere 7, Florence 1968, p. 30); "Error est dicere: 'sine Aristotele non fit theologus'; Immo theologus non fit nisi id fiat sine Aristotele" [It is false to say: 'without Aristotle you do not become a theologian'; rather you only become a theologian without Aristotle.] (Luther, Werke in Auswahl 5, ed. Vogelsang, 1955, p. 323). The criticism came from the 'scientific' as well as from the theological side. This is understandable, for in natural science the new objectivity of nature, which appears in the mathematical perspective, and in theology the new autonomy of the human subject, which constitutes the subjective principle of this perspective, is the source of the anti-metaphysical motive. Emancipation from the powers of nature and of tradition is the implicit aim of modernism. And Aristotelian metaphysics is felt to represent these powers in the world of thought. The anti-metaphysical attitude, progressive in those times, has become conservatism in our time, in which we suffer from the - technical as well as cultural - *success* of the above-mentioned emancipation. Besides, what then was said about the inaccessible and esoteric character of scholastic metaphysics, can now be said likewise about many highly abstract branches of mathematics, mathematical logic, philosophy of science, and theoretical physics.

and the above-mentioned 7 are rationalizations of it. Aristotelian metaphysics mainly had to be dethroned, not because it really ran counter to mathematical reflection or physical understanding, but because mathematical insight was felt to be its legitimate successor to the claim of most fundamental knowledge.

5.4 Learning the lesson

Now that we have dealt with all objections to metaphysics originating in the confusion of its proper aim with the form it takes under the influence of the mathematical paradigm, it is time to describe this proper aim and the possible methods corresponding to it.

Metaphysics is the discipline of the explicit realization of the theoretical openness of the human mind. Its method can be characterized by the third degree of reflection, as has already been explained in chapter 2. This implies that it has to transcend any previously given method. As the Dutch logician Evert Willem Beth put it: it is thinking bareheadedly. Of course this might be considered as a permit for methodological chaos, but it is not. For the hard and difficult rule of this level of thought is: reflect about what you are presently doing. To this rule one might on the other hand object that it prevents any result beforehand. This is not true, however. Knowledge of principles cannot be swept away by reflection. If we consider as an example the principle which I have called 'principle of structurability,' which makes mathematical thinking possible, we see clearly that however rigorously we reflect on our formulation of that principle, our conclusion can never be that mathematics is based on a misunderstanding and has to be abandoned as a branch of knowledge. It may be that our reflection has methodological consequences for mathematics or consequences for the contents of a philosophy of mathematics or technology, or even for the methodology of philosophy itself, but it can neither change the basic questions, nor settle them forever. These questions are only *clarified* by such reflection, which means that we learn to express more clearly what the principles are which we justly assume already implicitly in thought and action. Not that it is impossible to assume implicitly something wrong, but even wrong assumptions must have some foundation.

If the role of reflection in metaphysics is clarification, it is seen to be distinct from the primary source of knowledge. In modern philosophy there is a tendency to confuse reflection and intuition, and we should avoid this mistake. Intuition is often understood as some flash of insight, occurring now and then in our intellectual life. This is caused by a one sided interest in problem solving. If we consider a particular problem, its solution by intuition is a one-off event. But intuition is always present, for we cannot remember knowledge of principles by habit and external identifiers, as we do with conceptual knowledge. Principles are nothing if not actually understood. For instance what is traditionally named the principle of non-contradiction or the principle of identity, indicating the inner unity of being,[184] is leading us in all our thoughts, consciously or unconsciously. This does not mean that we cannot be inconsistent in our thoughts, but it does mean that we always strive for consistency, not in a pre-determined formal sense, but with regard to the actual content we want to express. When Hegel writes that it is contradiction which moves the world, this is precisely what he means. An apparent contradiction forces us to move towards a higher level of thought, not only personally, but also systematically and historically.[185]

We have learnt that principles are no formulas or recipes - that would be mathematism. Their formulations are the result of reflection, which can always be improved. Principles are understood before any formulation, before any reflection and before any theoretical expression. Of course their understanding is simultaneous with their actual practical use - also in the practice of thinking - in which they are active in a functional form. Identifying principles with their explicit expressions leads to the mistake of separating them from their contexts and from each other. They seem to be disparate and the problem then seems to be how to unite them into a metaphysical system. But this approach starts from the wrong side. If the principles really belong to the intelligible content of reality, they cannot be disparate, and if they do not belong to this intelligible content, they are not really principles. Therefore one has to start from an intuition of the coherence of experience and try to develop it into an explicit account of the relationships of the many perspectives in which we usually see it. The different principles underlying those perspectives can then be understood to be inseparable from their mutual relationships.

Such an attempt at development of the systematic coherence of principles within their concrete context is what Hegel called speculative dialectics. Although it tends to result in something like a 'system,' this result can never be more than a provisory stage of the investigation. The process is as essential

---

184   Everything is one' is an ambiguous sentence. It expresses the unity of each finite being separately as well as the unity of infinite being or the principle of being. The identifiability of both meanings is itself a metaphysical problem. The expression 'unity of being' is here used primarily in the first sense.

185   See: M. Wolff, *Der Begriff des Widerspruchs, eine Studie zur Dialektik Kants und Hegels*, 1981; Thomas Kesselring , *Die Produktivität der Antinomie, Hegel's Dialektik im lichte der genetischen Erkenntnistheorie und der formalen Logik*, 1984, and by the same author *Entwicklung und Widerspruch, ein Vergleich zwischen Piagets genetischer Erkenntnistheorie und Hegels Dialektik*, 1981.

as the result. It is the *motion* of thought, rather than its end-point, which clarifies the relational unity of principles. This is in accordance with the practical effect of principles in thought and action. They do not constitute specific contents, but they direct the *motion* of the process in a way comparable to Kant's regulative ideas.

Although the relational unity of the principles can probably be expressed most coherently in a dialectical philosophy, progress in the realm of *metaphysics* itself cannot be dialectical. The same content which has to be clarified always remains the same, and is already implicitly present. Maybe new principles will be discovered, but reflection will tell us that actually they were already known, only masked by others. Metaphysics essentially has the form of a *philosophia perennis* because its content is so fundamental that it is beyond all revolutions. This may sound conservative, but on the other hand the true revolutionaries are those who seriously try to realize the values of tradition. The ingredients of metaphysics are already present in the most ancient mythologies, and their historical metamorphoses reveal at once the progress of reflection and the constancy of content.[186]

It seems to be a heavy claim that human knowledge has a deep and constant basis, not to be overthrown by any philosophy or by any practice. But it cannot be otherwise, since the contrary either replaces knowledge by construction or restricts its content to historical fact. The first case leads us back to mathematism, the second to historicism. In both cases the relation between the act of knowing and its content is reduced to an abstract identity from which all tension is eliminated. The construction is as it is, because it is constructed to be so. The historical fact is as it is because it is conceived to have taken place at a certain time. All critical reflection is thus deprived of its object, and instead of liberating human thought from the regime of absolute principles this conception hands it over to uncritical acceptance of chance. Strictly speaking, with the reduction of the relationship between the act of knowing and the known content to abstract identity, the idea of human knowledge is actually eliminated. Such knowledge means an identity of content between distinct modes of being: reality and intellect. The distinction of intellect and reality is the origin of the tension which makes the project of understanding the world of our experience worthwhile. Mathematism,

---

[186] This coherence of content between myth and metaphysics does not provide a sufficient basis for theories which try to *reduce* one of both forms to the other, neither qua form of knowledge or qua social function.

historicism and scepticism eliminate it, by reducing the distinction either to an abstract identity or to a separation, which is equally abstract.[187]

If there were no real and known and therefore remaining principles, there would not be any human knowledge, and any theory would be futile. Metaphysics is the continuing project of clarifying these principles by reflection, and thereby clarifying the meaning of *all* human knowledge. This project is based on the third degree of reflection and is, therefore, not allowed to leave anything given in human experience out of consideration. It cannot abstract from the individual cases, nor from quality and change, nor from any human practice. This makes it the most difficult and ambitious project of the human mind. For there is no methodological instrument, except reflection, for distinguishing the relevant from the irrelevant, fact from fiction, reality from appearance. It is hard to be forced to admit that any idea may be relevant, even if it is not factually true. The objections against such an attitude, however, generally originate in an underestimation of the force of reflection. For reflection has to place all considerations first of all in the field of their possible contexts, among which all possible objections to the said considerations. So there is nothing uncritical or credulous about it, but it has a positive aim. In metaphysics this is the clarification of the principles. This aim guards it from scepticism. Metaphysical thought has to hover between scepticism and dogmatism, and it can never choose one of both as the easy way out. Neither a dogmatic system, nor nihilism, which is the adoption of a dogmatic scepticism. There is an archetypical situation in this, nowadays usually connected with the question of the 'final foundation of knowledge.' This expression, however, is a little bit misleading if it is used in connection with metaphysics, for it is not the *explicit* metaphysical knowledge of principles which is a foundation for other kinds of knowledge, but the knowledge of them which is *implicit* in all human practice.[188] Metaphysical thought aims at *investigating* these principles, not at establishing them as starting points of other disciplines.

Now one may wonder how it is possible to write a whole section about principles without actually mentioning them. As it is with dialectics, it is of not much use to talk about metaphysics as if it were something outside us, instead of our own activity. That would be 'external reflection,' as Hegel calls it. For reflection on principles such an external point of view would not be

---

187 Even in the case of absolute self-knowledge, as scholasticism ascribed to God, there cannot be said to be an *abstract* identity between knowing and content but this leads us too far from the subject.

188 This seems to me to be the point of Peircean as well as of Wittgensteinian criticism of metaphysics, and of the philosophical tradition in general.

adequate since, while reflecting, we use the principles all the time. They are neither external nor internal with respect to our thought, for they precede this distinction.

As an example of reflection on principles I shall discuss in the next section the principle or principles of structurability on the basis of what I have said in Chapters 1 and 2. In the last sections some problems inherent in the idea of metaphysical investigation will be discussed, including the problem of the unity and plurality of principles, which we shall encounter already in the next section.

5.5 The principle of structurability

In Chapter 1 we saw how the principle of structurability is related to Aristotle's principles of matter and form. It is responsible for the form of material being as such, which is the moment of indifference present in all material being. According to Aristotle the principle of matter itself is passive potency or pure determinability. But matter is never without form, so in a real material being there is never *pure* determinability. The determinability is *inherently* restricted by the form, which means that a physical being can change, but not arbitrarily. In Aristotelian physics the potency of change depends on the nature of the thing liable to change. In modern physics this means that it is subject to laws. But these laws are thought of as universal, not as the proper immanent laws of this specific individual being. Rather than depending on its nature, they *determine* its nature. Instead of thinking about determination as of something immanent in the concrete individual, modern physics conceives of it as in a sense external to this individual being, which is only an instance of it. This means that the individual is thought of as completely determined by external laws, and therefore as being itself nothing but pure determinability. Instead of being an intimate union of mutually permeating matter and form, the substance[189] of modern physics is an external union of pure individuality and ideal universal law.

It is mathematical objectivity which has been paradigmatic here, and as a consequence of this paradigm the laws naturally become quantitative. Hegel determined quantity as being which is indifferent to its own determination.[190]

The principle of this indifference, which was named *intelligible matter* by Aristotle, is what has been described in Chapters 1 and 2 as the principle of structurability. It encompasses determination and determinability, which could be named quantitative matter and quantitative form in an Aristotelian way, but which manifest themselves as discreteness and continuity, the principal moments of all quantity. Aristotle considered continuous and discrete quantity as two separate kinds of quantity, but in modern mathematical thought their essential relative union appears, which is admirably expressed by Hegel in his chapter on pure quantity in his *Science of Logic*.[191] Whereas physical

---

189   The difference between the Aristotelian and the modern concept of substance is a rather subtle one. Substance in the Aristotelian sense must be understood primarily in contradistinction to property, accident, activity: it is the relative autonomy of a finite being. The modern notion of substance is conceived in contradistinction to change or process: it is essentially the permanence of nature as a whole. A principle of substance can in modern science be said to underlie all conservation laws. The Aristotelian notion of substance also occurs in science, but in a more inofficial way within the *informal* ontology in which the validity of laws is understood as a property of *something*: e.g. particles, wave packets or fields, which can be of a very impermanent character. Both notions of substance have in common, that they contain a tension between an individual and a universal moment, but in the modern notion these moments are understood as much more widely separated.

190   See the opening sentences of the Chapter 'Quantity' in his *Science of Logic*, Werke 5, p 209: 'Quantity is pure being in which determination is no longer manifest as identical with being itself, but as sublated or indifferent.' and the parallel text in the *Encyclopedia I*, Werke 8, p 209: 'Quality is the first, immediate, determination, quantity is determination which has become indifferent to being; a border, which is not a border as well.'

191   Werke 5, p.211 ff. 'So continuity is simple identical self-relatedness, which is not interrupted by any border or exclusion, but not *immediate* unity, but unity of individual [für sich seiende] units. The mutual externality of plurality is still contained in this, but at the same time as indistinct, uninterrupted. Plurality is potentially [wie sie an sich ist] present in continuity; the many are one and the same, each equal to the other, and plurality therefore is simple indistinct equality. Continuity is this moment of *self-equality* of external being, self perpetuation of distinct units in what is distinguished from them. Therefore quantity in its continuity immediately possesses the moment of discreteness, - repulsion, which is now a moment of quantity. Contiguity is self-equality of the many, which do not yet become mutually exclusive; Repulsion expands self equality into continuity first. Discreteness, therefore, is on its part confluent discreteness, the units of which are not connected by emptiness, the negative [as it is in atomism], but their own contiguity, and

being has become more abstract in modern thought, mathematical being has become more concrete, which again illustrates the intimate alliance of physics and mathematics at the dawn of the modern era. The perspective of mathematical thought has become applied to a far more extensive domain in this era, and at the same time it has been developed internally by a process of mathematical self-reflection, which produced the refinements of modern mathematics and eventually the explicit forms of mathematical self-reflection: mathematical logic and information technology.

One might ask in this connection whether continuity and discreteness are again principles, contained in the principle of structurability. Or are they 'reflexions' of the more general principles of determination and determinability within the perspective constituted by the principle of structurability? We must be on our guard here for the inclination to systemize principles, because we can easily fall back into either the common-sense paradigm of scholastic reasoning or the mathematical paradigm of modern science. It has already been remarked that even Hegel did not quite succeed in reconciling the dynamic and the systematic aspects of his thought. In section 5.7 the problem of the unity and plurality of principles will be touched upon again. In a methodological perspective it seems to me that to develop the art of being systematic without building a system is the main task in developing future metaphysics. There is an interesting parallel with information technology here. Systems of information processing - even the so-called learning systems - cannot work without a preconceived program and hardware design, which may be universal with respect to those tasks which are mathematically comparable to the calculation of computable functions, but which is never universal with respect to interaction with its surroundings. There must be some basic structure which defines the interface through which information enters and leaves. If human intelligence is not reducible to such an information-processing system, it can only be because it has no pre-designed basic structure defining its interactions with the world. Only in that case can metaphysical investigation as distinct from mathematical science be regarded as meaningful. Otherwise the best we could hope for is finding the basic structure of our mind. This makes it clear that if metaphysical investigation is meaningful, the outcome can never be a system with a fixed structure. Yet we have to distinguish principles and to investigate their order and relationships. We have to make

---

this self-equality is not interrupted by these many units.' If one compares this difficult text with Cantor's notion of a set, one can be as amazed as when Hegel's development of matter from space and time is compared to the fundamentals of relativity theory. But this is no anticipation of future science. It is the other way round: science gradually incorporates more of its implicit principles into its theories. Speculative philosophers such as Hegel and Aristotle, however, tried to explain these principles in their own right.

a map of the various perspectives from which we can understand our experience and to evaluate their relative adequacy with respect to certain phenomena. Our aim, however, is not the creation of this map, but the attitude of reflecting and gaining insight.

5.6 The metaphysical attitude

Before the name 'metaphysics' was introduced there was Aristotle's book on 'first philosophy' and Plato's use of the word εποπτευω, which was taken from the language of the Greek mysteries and meaning something like 'to have a vision', 'to investigate', 'to contemplate'. In both cases the *search for what is most fundamental*, which was already the aim of presocratic philosophy from the very beginning, is intended. Later the meaning of the word 'metaphysics' has been identified either with some specific form or content of such a search, or with a caricature of such a form. Therefore I propose to return to the original intention and first to investigate what it *necessarily* involves.

The expression 'search for what is most fundamental' requires affirmation of several presuppositions in order to be meaningful:

1. There is an *order* of more and less fundamental contents.
2. This order has some kind of enveloping totality or maximum: the *most* fundamental.
3. This totality is not simply given: it requires a *search*.
4. We possess the *faculties* to undertake such a search.
5. There is a certain *value* in undertaking such a search.

In this section I shall discuss these presuppositions and their consequences in order to clarify this proposed use of the term 'metaphysics'.

Ad 1

This order we could try to identify with the order of *presupposition*. What is presupposed for the understanding of an expression of thought is not necessarily itself explicitly expressible or already so expressed. It is only implicitly affirmed by discussing the original expression. The present subject may also be taken as an example: In discussing the idea of metaphysics as the search for the most fundamental, we implicitly affirm the existence of an order of more and less fundamental contents. Now it is clear that we cannot open our mouths to say anything, without presupposing a lot of other things. Just as in a dictionary we cannot explain the meaning of a word without using other words. And clearly this does *not* imply that some words are more fundamental than others. So 'presupposition' is not sufficient to explain our 'order'. We mean: more fundamental presupposition. The presupposition that one can explain the meaning of a word with other words is, for instance, a fundamental presupposition of making and using dictionaries. But then our argument becomes circular. By 'more fundamental' we do not mean an arbitrary presupposition, but a *necessary* one. A word can be explained in many ways and its use, most of the time, does not require an explanation at all. That is what is wrong with the dictionary example. The order of the more or less fundamental acquires a transcendental nature: the more fundamental is a necessary condition for the very existence of a meaning of the less fundamental.

Ad 2

Mathematically it is not true that any order has a maximum, but it *is* true that any order has a principle of ordering. It has been argued in chapter 2 that all perspectives of understanding have such a principle. These principles are more fundamental in our sense than anything that could appear *within* the corresponding perspective. Therefore one could claim that metaphysics is the search for a principle of all principles. In a certain sense this is true, and Plato already expressed this immodest claim by talking about the idea of ideas. One should keep in mind, however, that this does not imply that we are able to know this first principle in the same way as we seem to know the restricted principles constituting the perspectives in which we act and think. From the antinomies of abstract mathematical set-theory one can learn that one cannot simply compare a principle and that which appears in the perspective constituted by it, within the same order. The 'set of all sets' cannot exist, for the principle of set-theory is not itself expressible as an object within the universe of sets. So the expression 'principle of principles' is only a metaphor pointing towards something *beyond* the realm of what we usually call principles. The same is true for terms such as 'maximum,' 'totality,' 'absolute,' etc. Yet these terms cannot be empty words, for any order presupposes a principle. It has often been tried to avoid this conclusion by introducing the idea of ordering by chance. The concept of chance, however, loses its sense if it is stretched so as to denote an attribute of the absolute. What happens by chance is by definition that which does *not follow* from some law or ordering principle. If the concept of chance is used to express something which amounts to the judgement that *everything* happens by chance, the distinction between chance and necessity has been lost. In the meantime a discussion about what is most fundamental has already been admitted as meaningful.

Ad 3

How can we search for something we know for sure we will never find? One could ask with equal right: how can one search for something which one has not to a certain extent already found? All search includes a tension between the found and the unfound, the known and the unknown. Such are all major human enterprises: science, technology, art, politics, ethics, religion. Why should metaphysics be an exception? Perhaps the comparison with technology is considered to be inadequate. Technical perfection seems to be within reach. But in fact technological progress presupposes continuous *im*perfection. Ask those engaged in software engineering! The ultimate aim of technology lies *beyond* technology, in the realm of human life, and will never be reached completely by technology. This is what the idea of *progress* is all about: its achievements do not constitute its ultimate meaning, but the openness with respect to future possibilities due to the fact that its ultimate aim lies beyond its achievements. This is what Aristotle already remarked about life: its end is not its aim!

Ad 4

If we do not have the faculties to search for the most fundamental, how do we know that we do not have them? Does not this knowledge involve knowledge about the most fundamental? And if we possess such knowledge, does not that prove that we have the faculties to possess it?

Ad 5

The question of the value of the search for the most fundamental is very difficult. But on what basis shall we discuss it? If we can find a definite basis for such a discussion, should we not discuss how fundamental it is? If we only can find a provisory basis, should we not look for a more fundamental one? Only for someone who has *found* the most fundamental would this not be a problem, but such a person would not be interested in the answer to the question whether it is valuable to search for it. In fact one can not justify such a search without already claiming certain results of it. One simply undertakes it or not: "Go forward and faith will follow you."[192] Only afterward can the value be judged. Of course one might be worried that later results can annihilate established values, but then it may also be the case that for values the same holds as for principles: their number and their relationships may change, but once discovered their validity remains.

---

192   "Allez en avant et la foi vous viendra" This 'bon mot' is ascribed to Laplace and is supposed to concern mathematics.

One might wonder who would dare to hamper such an open minded investigation. Yet it is easy to caricature it or to depict it as useless. The motive for this invariably is the silent conviction that one already knows the answer to the question what is most fundamental. It is clear that in that case one might be interested in avoiding the risk of this conviction being shaken by any investigation whatsoever. Curiously enough I cannot think of any other motive to object to metaphysical investigation in this sense. Nevertheless someone could agree that there is nothing against it, but that it has not been shown to be of any use either. Is it not possible to keep human knowledge open-ended *without* designing an explicit discipline for that purpose?

I have to confess that I mistrust this attitude. For why should the tendency in all human investigation to make everything explicit and thematic suddenly stop short before the theme of the most fundamental? Could not mathematical logic, foundational research in physics, etc. be rejected on the same grounds? Yes, indeed, these subjects actually *are* considered as somewhat useless by most 'working' mathematicians and physicists. And the same psychological motive seems to lie behind that opinion: it is threatening to confess that one has to *investigate* the foundations, for that means that they are not as certain as one would want them to be. Socrates' fate is always imminent for the philosopher.

5.7 Metaphysics as the search for the principle of principles

Philosophy has to take its starting point in life itself. It has to disentangle all the perspectives from which we conceptualize and reconstruct experience, analyze the way we talk and think about it and describe the essence of the phenomena we express in this talking and thinking in relation to the principles we silently presuppose. We have to ponder the one-sidedness of the perspectives, the abstractness of reconstructions and the relativity of essences, but without philosophical arrogance, for the vast plurality of perspectives and reconstructions already compensates for their one-sidedness. Therefore we should look for systematic coherence without premature attempts to construct a system. By this gigantic labour, which Hegel in his youth called '*realphilosophie*,' the philosophy of reality, can we hope to discover a plurality of real principles. I have given the example of the principle of structurability. We recognize these principles by their unavoidable and at the same time perplexing character. Without them it turns out to be impossible to make sense of the perspectives we adopt, and use to order experience, both in a theoretical and in a practical context. These principles cannot be understood as 'abstract' in the sense that they can be considered apart from the context of real experience. Mathematical structures can, for example, be *considered* without connection with the observable world, but they cannot be understood to make any *sense* without that connection. Their sense comes from the principle of structurability and that is precisely what constitutes the *connection* between the abstract mathematical world and the observable world. The principle, therefore cannot even be considered without taking into account the totality of experience in its pre-conceptual immediacy.

At another level of philosophizing, we can try to investigate the *inner coherence* of the principles, either as principles of being or as principles of knowing. Plato's doctrine of ideas, Aristotle's *Metaphysics*, Kant's *Critique of Pure Reason*, and Hegel's *Science of Logic* are outstanding examples of such attempts, which eventually lead us into the realm of metaphysics. The idea of an inner coherence of principles obviously raises the question of a principle of principles. This also follows from the fact that philosophizing is a human practice like any other, and therefore presupposes a perspective of its own. As this perspective has been characterized as the third degree of reflection, which deals with principles, the principle constituting the meaning of this perspective must be the principle of principles. Probably this is the reason why the question what philosophy is, seems to be equivalent to an ultimate metaphysical question.

Perhaps a few words should be said about possible problems with the reflexive character of the formula: 'principle of principles,' which might remind us of the 'set of all sets,' and which could therefore be paradoxical in itself. But even in mathematics such formula's are not always paradoxical. In abstract Algebra the notion of the category of categories[193] gives no problem. It is only when a mathematical concept is regarded as an externally limited extensional domain that the problems arise. One can talk without contradiction about e.g. the concept of 'concept' etc.[194] Only when the content of a concept is represented by an object in the order of its extension, *and* this extension is conceived of as an externally limited mathematical plurality, a contradiction arises. The reason for this is clear: it cannot meaningfully be asked on which side of an externally defined limit lies that by which the limit is defined. So the interpretation of a mathematical concept, such as 'set' as itself denoting a certain mathematical *object*, an externally

---

193 A category is an algebraic structure representing a domain of mathematical objects together with the appropriate mappings between them. See e.g. the article 'categories and functors' (53) in the *Encyclopedic Dictionary of Mathematics* MIT press 1980.

194 See for a logical treatment of these problems: Koyré A., *Epiménide le menteur*, s.d.

limited extension, leads to a contradiction. If we refrain, on the other hand, from such an interpretation, for instance in the way John von Neumann introduced the notion of class in set-theory, the strict univocity of the term is lost. A class can be either a set or a proper class, which is by no means the same. Analogy is introduced here, which is normal in ordinary and philosophical use of language, but exceptional in mathematics. In other mathematical cases in which reflexivity or self-application does not give a contradiction, it can be shown that some form of analogy is involved. We shall see that the expression 'principle of principles' involves analogy too.

The formula 'principle of principles' has its reflexivity in common with traditional formulations such as: the idea of ideas; being as being; νοησις νοησεως. In chapter 3 sections 2 and 3 I introduced some considerations concerning the function of such reduplicative formulations. I compared them to the mathematical form of self-application of functions,[195] expressing in mathematical thinking the external representation of the subject's act of external self-representation itself. In modern philosophy it acquires the meaning of a self expression of the subject as *causa sui*, as the *origin* of its self expressions. This has led me to the conviction that the ultimate meaning of such expressions is *originality* itself. Now that is already what I mean by the word 'principle': that which is first with respect to a certain perspective, and therefore in a logical - not a temporal - sense the origin of the perspective. The principle of principles, therefore, does not only intend the origin of *all* principles, but also of *principality* itself. What does this mean?

If one is convinced, as I am, that human knowledge is never direct and complete, which means that it is capable of grasping reality intuitively only under a certain perspective and that it has to reconstruct its meaning from many such partial intuitions, one has to admit that such a capability cannot itself be based on a limited perspective. The notion of 'principle' does not make sense unless the principles are supposed to be real and intelligible. So

---

195   In the current form of mathematical set theory it is not allowed to apply a function f to itself as an argument. An expression such as f(f) is meaningless there. In purely functional algebras, such as Alonzo Church's Lambda calculus and Haskell B. Curry's combinatory logic, there is no Ôbjection to self-application. Functions are understood there as operational entities, comparable to computer programs. A computer program too can be applied to itself as an input file, something which regularly happens with programs testing a stock of stored programs for viruses. Dana Scott constructed a set-theoretical model for such purely functional systems, in which also a set-theoretical equivalent for self application can be meaningfully defined. Nowadays self-application is understood to play a central role in recursion-theory as well as in all applications of Cantor's diagonal method, such as Gödel's incompleteness theorems.

'principality' includes the intelligibility of reality. This involves many problems, of which I mention two particularly important ones:

1. The 'principle of principles' must also be the principle of intelligibility. Then, however, it can no longer be a *restricted* principle, like the ordinary principles. But why call such an 'unrestricted' principle a 'principle' then? This can no longer be univocal denomination, it must be a case of analogy. But what is the basis of this analogy?

2. And what with the *plurality* of principles? Is this a plurality *only for us*, caused exclusively by the restricted nature of our interest? Is this conception consistent with the demand that the principles must be *real*? But is their plurality as real as they are themselves, so to speak by grace with respect to restricted intellects in order to enable them to understand something of reality. How then, can this real plurality exist within the domain of the unique 'principle of principles'?

Both problems are classical. The first is the problem of the limit of human understanding, with which all great philosophers, from Plato to Wittgenstein have wrestled. The second is the problem of unity and plurality, which is even older, and which is actualized in our time by philosophers such as Lévinas, Liotard, Derrida, who all tend to reverse the traditional primacy of unity and to postulate a primacy of plurality. Such problems clearly are not 'solvable' in the sense in which technical, conceptual or mathematical problems can be. One can at best determine one's *position* within the realm of these questions, and hope to thereby to contribute to their clarification. It is clear from the preceding chapters that I am obliged to do so, in order to give a plausible foundation to my theory of principles.

First I shall try to determine my position towards each question separately, then a connection between the two will be established.

1. The question of intelligibility

My position will be characterized by an attempt towards a rehabilitation of the *receptive* element in human knowledge.

It has to be admitted that the conceptualization and reconstruction of experience in culture and science are to a large extent human products. This does not imply, however, that they are completely arbitrary. They are *meaningful* and their meaning depends upon principles, which are *not* human products but ideal realities, enabling the perspectives under which we can meaningfully conceptualize and reconstruct the world of our experience. At least implicit knowledge of principles must have been present from the very beginnings of culture and science. This implicit knowledge is the basis of philosophy, understood as the activity of trying to make this knowledge more explicit. Culture and science, therefore, are not founded upon philosophy, nor the other way round, but philosophy has the same foundation as they have: intuitive knowledge of principles. Even with respect to this knowledge, be it implicit or explicit, there must be an active contribution from our side: abstraction or the distinction of the essential and the accidental. But this distinction cannot be a construction or an artefact of any kind, for then it would lose its sense. It has to be the result of a *receptive* intuition for a possible meaningful order of the totality of our experience. This intuition can be shared through recognition of the meaningful character of the resulting order. Individuals can grow up in a culture and grasp its meaning. This cannot be mere indoctrination, for indoctrination already *presupposes* a culture. Adults can become interested in a branch of science and grasp its basic ideas. This could be construed to be indoctrination if it were not the case that such individuals have in fact *created* the whole body of scientific knowledge as a meaningful and fertile perspective. Persons can also get interested in philosophy and in the process of developing it, they gradually surpass any established doctrine. I do not believe that these facts can be explained by pure chance, but even such an explanation would presuppose a real principle. Therefore I hold that principles must penetrate into our minds because we possess receptivity for them. We can even understand by philosophical reflection *that* this must be the case. This, however, is in itself only a *conclusion* by reasoning, and the intuitions on which it is based remain implicit. Yet, we know that it must be based on intuitions if it is true. These intuitions cannot have the nature of receptivity with respect to a single principle at a time, otherwise we could conclude nothing about the general nature of knowledge of principles, not even about the alleged impossibility of such knowledge. Our intuition, therefore, must have a wider horizon than the particular principle we are grasping on a certain occasion. There must at least be an implicit understanding of principles *as such*. This is what I call intuition of principality, or of the principle of principles. It is very probable that this intuition too, besides the intuitions of particular principles, must be constitutive for culture and science. Philosophy cannot monopolize it. The task of philosophy is to make our knowledge of it more explicit, and in so far as philosophy has explicitly *made* this its task, it is called metaphysics.

Metaphysics, therefore, has to dive into the deepest levels of our experience, an activity which, if it were undertaken with the exclusive aim of enriching our existence, would be called mysticism. The aim of metaphysics, however, is explicit understanding. In existential depth, it is therefore inferior to mysticism, but in clearness and generality of the result it should - by its own standards - be superior. This is precisely what characterizes its limit: where it becomes unclear and mystical, one approaches the limit; which does not exclude that next time clearness can be carried on a little further.

The 'principle of principles' thus certainly has a status different from the 'normal' principles. It does not constitute a restricted perspective, and it cannot be made as explicit as principles which do constitute such a perspective. It is denoted in an *analogous* way by the term 'principle of principles' and that is not surprising if we consider what we want to express by that term.

2. The question of plurality

It will be clear that the plurality of principles cannot be a numerical, quantitative or mathematical plurality. A principle can by no means be understood as an individual object or as a constitutive part of a quantitative whole. It exists only in so far as it is active in some real being.[196]

Principles, then, must be united inside the reality of which they are principles. Therefore we can conclude that their complete union must be being as such, or, in Aristotle's words, being as being. In neo-Platonism this has been denied. It was argued, that nothing else than *unity* as such, which was supposed to be 'above' being, could unite the principles. But if it is not clarified what is meant by such terms as 'being' and 'unity,' one runs the risk of struggling over mere words. In order to clarify this, the question of *how* then principles can be united at all, should be dealt with.

The whole notion of uniting principles points towards Hegelian dialectics. There we find switches between opposed perspectives, resulting in sublation of the opposition by showing that the perspectives are based upon opposite *moments* which are correlatives of one and the same relationship. This relationship, however, may also presuppose a certain perspective which evokes its counterpart, and the dialectical development proceeds further.[197] Of

---

196   Aristotle's criticism of Plato's doctrine of ideas in the 13th and 14th book of the Metaphysics stresses this point emphatically.

197   As an example we can mention the development of the notion of free will in the preface of the *Philosophy of Right*: a) In order to be free the will cannot be bound to any content b) In order to exert its freedom it has to bind itself to a certain content c) Its freedom precisely consists in being bound *by itself*. ca) The contents of will which fist present themselves are those that lie in human nature. Freedom then is to follow one's nature. cb) One cannot in everything follow one's nature, for it dictates unreconcilable deeds and attitudes between which one has to choose. Therefore freedom is the power of arbitrary choice. cc) Purely arbitrary choice has no value, it is sheer indifference and therefore below the dignity of freedom, which presupposes value and is itself a value. Therefore a choice can only be free if it is also a choice for the realization of freedom itself.

course, as it has become clear in the earlier chapters, there is more to Hegelian dialectics. But the idea I borrow from it here is only that principles can be united by *relative opposition*.[198] This implies that unity and plurality are themselves principles, connected in this intimate way. They are ultimately only relatively opposed, which means that in an absolute sense they are one, that is, one and the same principle.

Now what do we mean by 'unity' and 'plurality?' Probably this depends on the form of reflection in which we think about it. Therefore it is useful to investigate the forms of unity and plurality as we understand them from the perspectives of the first and second degree of reflection.

In reflection of the first degree the fundamental manifestation of unity is the unity of a *thing* or finite substance. The perspective in which there are finite substances is presupposed in the conceptualization of experience. It is *something*, an individual reality that can be subsumed under a certain concept by a judgement of which it is the real subject. If there are no 'somethings' the concepts lose their sense for they cannot be applied to anything. *What* exactly is postulated as an individual substance depends of course on the context. What is in one context a property or an activity of a substance, such as colour or motion, may in another context be regarded as itself a substance, such as ultramarine or a hurricane. We do not escape, within the context of conceptual thinking, from the necessity to postulate substances as that which we are thinking about. The unity of a finite substance is essential for the applicability of concepts. Two meanings of plurality are linked to this sense of unity: the external sense: plurality of substances; and the internal sense: the plurality of manifestations (properties, activities) of one and the same substance.

---

198 For the notion of relative opposition see chapter 1, section 4. The metaphysical use of this notion has been introduced explicitly by the Dutch philosopher B.A.M. Barendse in his essay on intersubjectivity and corporality. As a source he mentions Thomas Aquinas' theological doctrine of trinity, in which the notion of relation was given a transcendental meaning. It was detached from the Aristotelian restriction to a category, and used in a sense in which subsistence and relativity were ontologically identical, although logically distinct. This doctrine also influenced Jacob Böhme and along this line also Hegel. Barendse influenced a number of neo-scholastic Dutch thinkers, of whom P. de Bruin, J.H.A. Hollak explicitly investigated the link with Hegelianism. The literature is almost entirely in the Dutch language. J. Aertsen's article is provided with an English summary: Barendse B A M, Intersubjectief verkeer en lichamelijkheid, in: Barendse B A M, *Zich door het leven heendenken*, 1982; Hollak J.H.A., Wijsgerige reflecties over de scheppingsidee, 1975 ; Velthoven Th. van, *Ontvangen als intersubjectieve act*, 1980. Also in: Th. van Velthoven, *De intersubjectiviteit van het zijn*, Kampen 1988; Munnik René, *Van grens-verleggen naar ruimte scheppen*, over het raakvlak van metafysisk en techniek, 1992; Aertsen Jan A., Denken van de eenheid, 1990 pp.399-420.

In mathematical thinking we must regard structures as well as their components as being quasi-substances. Otherwise mathematics would be robbed of the tool of conceptual thinking. It would be reduced completely to the activity of mental construction which L.E.J. Brouwer took for its essence, but which in fact would lock up the mathematician in his ideal universe. Mathematical objects are *quasi*-substances, for their unity is somewhat different from the unity of finite substance in the first degree of reflection. In the mathematical realm the unity is *purely* and explicitly postulated, whereas in the realm of normal conceptual thinking it is postulated as being an underlying reality. The unity of a real finite substance is understood as an *inner* unity,[199] whereas the unity of a mathematical object is purely external. It is only by *mathematical* reflection on the conceptualized world, that we are able to think of finite substances as also postulated. For common sense the substances are simply there, we only name them. In mathematical thought we also find the remnants of the two senses of plurality of common sense. The plurality of structures and the plurality of elements of one and the same structure, but they actually amount to the same here and can only formally be distinguished. Moreover they are both external pluralities.

We have to admit now that the perspectives of unity and plurality enclose the perspectives of *inner* and of *external* unity and plurality, without committing ourselves to an ontological position with respect to the question whether any of the two applies to anything in an *absolute* sense.

The most obvious paradigm of inner unity and plurality is implied by the notion of a *person*. It refers to our own experience of remaining the same person through the plurality of all our experiences, usually expressed in the personal pronoun 'I'. But also to the idea of personal identity implied in the other personal pronouns. Thereby we understand the kind of plurality corresponding to this form of unity: a plurality of inner activities on the one hand and of interacting unities -but now no longer in a completely external sense - on the other.[200] I and you and he and we and they. If we come to

---

199   This was the source of Leibniz's difficulty in criticizing Descartes and Spinoza simultaneously. He saw clearly that extension could not be a principle of substantiality (*Discours de Métaphysique*, XII), but he did not accept Spinoza's monism either. It was probably from this difficulty that his monadology originated. It is remarkable that this connection is so fundamental that any philosophy or theory which attempts to save the mathematism of modern thought as well as the idea of finite substance, must necessarily end up with a kind of monadology. As examples Theillard de Chardin's theory of evolution and Rupert Shelldrake's theory of morphogenetic fields may be mentioned.

200   At quite another, more fundamental, level of understanding, the plurality of persons is analogous to the plurality of principles. In their intersubjective contact, persons do not appear as entities or substances existing side by side with each other. *I* as a continuous

think of it, this is how contemporary physics understands its world too. Mathematical subtlety has never succeeded in eliminating the notion of a *particle* from the vocabulary of physics, although with respect to the mathematical discourse this is an alien and even unscientific notion. But physics is not pure mathematics. It has to conceptualize a *physical* objectivity, and in such a conceptualized physical world, if there is action, necessarily there is something that acts and many things which interact. This 'conceptual scheme' of a plurality of interacting unities, which is applied to unities we reason about at whatever level, belongs to the essence of the perspective of the first degree of reflection. Without it, this perspective would lose its sense, and without this perspective physics would lose its sense. In physics nowadays, however, the particles can no longer be separated from their interactions. The more the moment of externality is adequately expressed by the mathematical formalism, the more the internal moment can reappear on the side of physical objectivity.

In the mathematical perspective, Cantor's notion of a *set* can be taken as paradigmatic. A set is a unity of unities. No interaction but external unification. Unity and plurality are already relatively opposed here, but the relationship is itself external with respect to what it relates, it is outside mathematical objectivity, performed only by mathematical thinking in postulating this objectivity. The content of the membership relation of set and element (the ε-relation) is determined by theoretical postulation only. The 'unity of unity and plurality,' which in the first degree appears in the form of interaction *between* individual substances and of a plurality of activities *within* them, here appears as the nature of the (quasi-)substance itself. The set *is* a unity of unity and plurality, that is its very definition. Only, whereas interaction is understood as founded in the nature of the substances themselves, the mathematical unity is understood as founded in external postulation. Therefore the unity of mathematical objects is only *quasi*-substantial and their plurality is external plurality.

Now the inner coherence of principles, as it is understood in the third degree of reflection, should constitute a relative opposition of unity and plurality, which is at the same time an *inner* unity. The principles are *by their own nature*, not by external postulation, a unity of unity and plurality. Each principle is a principle of many principles, not indifferently, as a set is a set

---

activity of self-identification, meet *you*, not as some entity in the world, but as the relative opposite of my..self in this process. The process is not confined to a factual individual. The subject is 'dialogical' from the very beginning, and this dialogue is presupposed in all experience, and therefore also precedes the world of experience as it appears in any of the degrees of reflection. This aspect of intersubjectivity will be discussed further in the next section of this chapter.

of many sets, but as a result of *self*-differentiation which is already immanent in its nature.[201] For the same reason all principles are essentially one principle. Each principle is in this particular way the principle of principles. It is useless to measure the plurality of principles by quantity, to say that there are finitely or infinitely many. Such considerations belong to the mathematical level, from which explicit consideration of principles is impossible. One cannot count principles, one can only *move* reflectingly through the perspectives they constitute. This is a city of which no complete map is available, in other words: all philosophical systems contain a certain amount of exaggeration which introduces a somewhat misleading mathematical component: the *structure* of the system, in which a certain *number* of principles is connected systematically. Far from being a philosophical tool, formalisation of philosophical notions is always leading *away* from the proper subject matter of philosophy. Of course a human philosophical enterprise necessarily presents itself with a certain structure, but this structure is 'the accidental by which the sage should not be misled.' That was probably the reason for Spinoza's remark: "I do not pretend that my philosophy is the true philosophy, but I do pretend that I *know* the true philosophy." This, and his attempt to present his philosophy '*more geometrico*' shows that he must have been aware of the problem. Philosophical exactness, however, is in a certain sense the opposite of mathematical exactness. It does not consist of a rigorous system, but of a systematic motion of the reflecting mind. Hegel's *Selbstbewegung des Begriffs*, self-motion of the Notion, is, therefore, closer to the true spirit of philosophy.

Speculative dialectics can be understood as the following of a philosophical trace. A principle is grasped intuitively, and developed by unfolding other principles participating in it as the moments determining any reality as seen in the perspective of the original principle. My attempts in this book to unfold the characteristics of mathematical reflection from the principle of structurability can be taken as an example. J.H.A. Hollak's attempt to develop the fundamental forms and moments of technology from his intuitive notion of the 'technical idea' (chapter 1, section) is another example. Marx's attempt at a theory of capitalism was not consistently dialectical, and perhaps it could not be because it is doubtful whether capitalism has a genuine principle or whether it is only based on a historical coincidence. The most comprehensive attempt at dialectical development of a principle is Hegel's philosophy. By some[202] this principle, the absolute idea, is interpreted as the principle of human individual existence. In any case Hegel's philosophy claims that it encompasses the totality of the human world, including the experience

---

201 Leibniz's monadology precisely lacks the idea of *self-externalization*. The monads are only *immanent* centres of force. Therefore, in spite of his mathematism, Leibniz has a problem with the externality of space and time. The monads as substances lack real subjectivity as much as Spinoza's unique substance. Therefore Hegel's philosophy, as an attempt to think substance *as* subjectivity, tries to correct Leibniz as well as Spinoza.

202 The idea was mentioned to me independently by Joseph Simon and by Jan Hollak.

of consciousness (Phenomenology of Spirit), the fundamental categories of thought (The Science of Logic), Knowledge of Nature, Subjective and Objective Spirit (The Encyclopedic System), and History (Philosophy of World History and History of Philosophy). Perhaps the human world can be seen in other perspectives, probably at least parts of it can, for instance in the perspectives of the positive sciences, including natural science, sociology, and psychology. I do not exclude that Hegel's system can be improved so as to encompass these perspectives too, or that an intricate texture of such systems is more adequate. A dialectical development, however, can include more or less of our experience, but not everything. Just as physical theories will always presuppose the determination of boundary conditions in order to be applicable, so a dialectical philosophy will always need a world of experience richer than itself in order to be meaningful. This must be so, because it is our history and experience which selects which determinate principles we are capable of making explicit in philosophical reflection.

Even in Hegel's dialectical philosophy the structure of the system has misled his interpreters, and perhaps even himself. Yet I maintain that the coherence of the reflective motion is a criterion of philosophical exactness, at least as stern as the criterion of consistency in mathematical theory. Neither of these criteria is of a purely 'formal' nature. It is true that a mathematical theory may be consistent by accident, but to *know* that it is consistent requires a clear intuitive content, which is relatively rare. The coherence of the philosophical reflective motion requires a genial intuition, which is even more rare. But if it is present, it is usually noticed by posterity.[203] Why else should we call the great philosophers great, although we probably reject many of their conclusions?

Now it is very probable that this is as far as we can get. Philosophy cannot explicitly consider principles unless they are already implicitly known within the culture and the era in which it operates. It must find them and test them by explicit philosophical reflection with respect to the actual world of experience. Hegel therefore described its task as 'Grasping one's time in thoughts.' Although a definite structure, expressing and connecting a certain number of principles and their corresponding perspectives, can lead away from the aim of philosophizing, any philosophy must necessarily have one. It is precisely this structure which reflects its historical limitation. What can be made explicit in a certain historical situation has an 'inner logic' because it reflects in a certain imperfect manner the absolute unity of the principle of principles. But in spite of this inner logic, it can never express adequately "God's thoughts before the creation of the world" as Hegel formulated it in

---

203 Some doubt whether genius in art, science and philosophy is *always* recognized is certainly justified, but to deny that permanent fame is a sign of it, is to deny its existence at all.

all modesty.[204] And this inadequacy is embodied in its definite structure, which it cannot do without. But all true philosophy is open to the idea of its fundamental inadequacy, and attempts to express it. This is its metaphysical side, which does not hamper its task of understanding as clearly as possible those principles which can be seen to be active in its world, but on the contrary furthers it. For it is the same understanding by which we know the coherence of those principles and the limitations of our expression of it.

What we can say about the way the principle of principles unites all possible principles is therefore mostly *ex negativo*. It can contain relative distinctions and oppositions, but it is not divisible by them either in reality nor in thought. It is the source of all necessary connections of principles, but it is not reducible to a system or structure. We cannot think of it otherwise than as itself an immanent principle which 'makes things as they are,' and encompasses everything we can 'dream of in our philosophy.' It may be objected that such a principle does not explain anything, but that is not true. It explains the possibility of principles, and therefore of understanding things in many perspectives. That is the only way, as far as I know, to reconcile the relativity of the way *in which* we know things with the absoluteness of the experience *that* we know them.

5.8 Principles and intersubjectivity

The fact that as an expression of the principle of principles philosophy must be tentative, but on the other hand must also try to express the reason for this in the form of its metaphysical component, presupposes that in all times there must have been a basis in experience for the form of this expression. A domain of experience which has attracted attention in this respect especially in our century is the experience of intersubjectivity. In all older philosophy the intersubjectivity of human experience has been taken for granted. The 'I' in modern philosophy is everyone. Even solipsism is naive in this respect, as is illustrated by the well known joke of someone saying: "I am a solipsist, and I do not understand why everyone does not share my conviction." Hegel explicitly mentions intersubjectivity as a necessary moment of subjectivity as such: Self-consciousness is a self-consciousness *for* a self-

---

204 The development of Hegel's logic out of Schelling's early philosophy of nature shows that this expression does not intend to 'look into God's playing cards', but to express the idea of a logic of creation. See: Wolfgang Neuser, Einfluß der Schellingschen Naturphilosophie auf die Systembildung bei Hegel: Selbstorganisation versus rekursive Logik, 1993

consciousness. *I* is *we*, and *we* is *I*.[205] The spirit is essentially intersubjective and this is precisely its substantiality. This is no new philosophical point of view, but it makes explicit, what has been implicitly presupposed in all of modern philosophy. By making it explicit, however, it is introduced as a theme of thought. In our century Peirce, Husserl, Buber, Rosenstock, Sartre, Lévinas, Girard, and many others[206] have elaborated on this theme in very original ways. One thing has become clear from this: the very fundamental nature of the theme. It is certainly not about a plurality of factual persons. It is about very deep presuppositions of all of our knowing and acting. A person is, as Aristotle already noticed, προς 'ολον - in a certain sense everything. In other words, a person constitutes a perspective on all being. Curiously enough it has this characteristic in common with a principle. There is a certain analogy between philosophy, as the activity of dealing with the unity and plurality of principles, and social life, as the activity of dealing with the unity and plurality of persons. The philosophical paradox that we need structure although it tends to lead us away from our aims, has its social counterpart. For we certainly need social structure too and it also tends to lead us away from the aims of social life. In both there is a continuous process of integration and disintegration, of opposition and reconciliation, of totalitarization and revolt. And we hope for progress in both, although we often have doubts as to its possibility.

The most important similarity between intersubjectivity and principles is the relational character of their unity. Eventually it is impossible to adopt

---

205 *The Phenomenology of Spirit*, IV The truth of self-certainty. *Phänomenologie des Geistes*, ed. Hoffmeister 1952, p. 140. "Es ist ein *Selbstbewusstsein für ein Selbstbewusstsein*. Erst hiedurch ist es in der Tat; denn erst hierin wird für es die Einheid seiner Selbst in seinem Anderssein; *Ich*, das der Gegenstand seines Begriffs ist, ist in der Tat nicht *Gegenstand*; [.....] Indem ein Selbstbewusstsein der Gegenstand ist, ist er ebensowohl ich wie Gegenstand. - Hiemit ist schion der Begriff *des Geistes* für uns vorhanden. Was für das Bewusstsein weiter wird, ist die Erfahrung, was der Geist ist, diese absolute Substanz, welche in der volkommenen Freiheit und Selbständigkeit ihres Gegensatzes, nämlich verschiedener für sich seiender Bewusstseine. die Einheit derselben ist: *Ich*, das *Wir*, und *Wir*, das *Ich* ist." [It (self-consciousness) is a self-consciousness before a self-consciousness. Only hereby it really exists, for in this its unity becomes eventually present for it in its otherness. I, which is the object of its notion, is in fact not *object*. [....] If a self-consciousness is objective, it is an I as well as an object. - In this the notion of *spirit* is already present for us. What presents itself further before consciousness itself, is the experience of what spirit is, this absolute substance, which is one with itself in the complete liberty of its internal opposition, which consists of different individual consciousnesses: *I* which is *we*, and *we*, which is *I*.]

206 In the Dutch philosophical tradition of this century B.A.M. Barendse's essay on intersubjectivity and the mediation of the body has been influential. As far as I know Barendse was the first to apply the scholastic doctrine of relative opposition to human intersubjectivity. See note 199 for references.

an external point of view. You *are* a certain person and you understand things from a certain perspective, even if you try to unite or reconcile persons or perspectives. Nobody is impartial and the best we can do is to try to understand precisely this. Yet, in both cases, we *know* that there is a unity, deeper than the ones we want to produce. If we were not certain about that, we could not even believe in our attempts at unification. We cannot, however be very sure about the nature of this deeper unity. We do not have any other method for identifying or distinguishing the unity of principles and the unity of persons, nor for comparing them to the god or gods of any religion, than by investigating certain characteristics of the place they occupy in thought and practice. Such unity(ies) is(are) certainly transcendent with respect to our explicit knowledge. Yet it (they) must be immanent in our theoretical and practical activities. The former presupposes the relating of perspectives so as to produce concrete knowledge and avoid becoming caught within one point of view. The latter involves the relating of persons in order to act as a concrete social subject and avoid being caught within an autistic delusion. Both forms of relating presuppose the fundamental correlativity mentioned above. However sceptical and cynical one may want to be, as long as one *expresses* such opinions and perhaps even recommends them, one proves thereby that one in fact believes in a fundamental contact between opposite perspectives and between persons with opposed attitudes. The crux here is, that the contact is not present *in spite of* the opposition, but that it consists primarily in the opposition itself. It is, for instance, unwise to try to solve the conflict between diametrically opposed philosophical points of view by first looking for 'common ground' before the nature of their opposition has been investigated. For it may well be that this nature will reveal the common ground as being precisely the cause of the conflict. So in human affairs it is equally unwise to cover up a conflict by suppressing feelings of anger or hatred and make a show of reconciliation on the common ground of compromise. In that case the conflict can burst out again in all its fierceness unexpectedly. It is within the very material of the conflict that the contact between the parties has to be looked for. When they become fully aware of what the conflict is actually about, there is a basis for a realistic solution. The germ of peace lies within the conflict itself.[207] In the philosophical realm, this of course reminds us of speculative dialectics, in which the proper insight comes from a 'sublation'

---

207   When one comes to think of it, this idea connects conceptions so far apart as Lévinas notion of the absolute Other, breaking into the sphere of Self and shaking it awake ethically on the one hand (*Totalité et Infini*, part III: Le visage et l'extériorité), and Heraklite's "War is the father of all and the king of all" and: "The inimical coming together and from the separating the most beautiful harmony" (Diehls&Kranz 22B53 and 22B8)

of the opposition between perspectives, which means from an understanding of the relative nature of the opposition.

Intersubjectivity brings the 'principle of principles' closer to our world of experience, makes it less esoteric. Metaphysics as the moment of self-reflection of philosophical thinking seems to mark an extreme possibility or limit of human reason. But the questions dealt with are at the same time closest to real life. Many a philosopher has wondered why we are so very sure that our fellow men exist in the same way as we experience existence ourselves. One may try to undermine this certainty by ingenious reasonings about 'human machines,' but we feel that they are beside the point. Certainly in all times there have been people who were considered as 'sub-human,' and the imaginary experience of the demasqué of a machine we took for a human being is not so very different from the experiences, described in many operas, of a gentleman being damasked as a slave or a valet. Such experiences *presuppose* real intersubjectivity, which remains completely certain. Of course in general the possibility of being deceived, only affirms our claim of being able to know the truth. It needs no explanation that thought experiments about machine-men presuppose real intersubjectivity too.

If we really succeed in deconstructing mathematism, it becomes clear to us that philosophy cannot aim at the construction of some kind of model or world-picture. Its only task is to clarify explicitly what we know and do already all the time. And we also realize that what is most common and most certain in ordinary life, is the most difficult to clarify. Professor Henk Oldewelt's[208] adagium: "We know much more than we think we know, only we know it much less clearly than we think we do," precisely describes the situation from which philosophical reflection has to depart. The 'dimension' of experience to which it applies implicitly presupposes already reflection, as was explained in chapter 2, but at the same time it is the most intimate experience we have: the experience of the reality of ourselves, our natural and social world and of other persons. It is the most difficult skill involved in the profession of a philosopher to remain close to this intimate experience,[209]

---

208 H.M.J. Oldewelt was a professor of philosophy at Amsterdam Universitiy from 1946 till 1967. Although his favourite form of philosophical expression was oral, he left many publications, all in the Dutch language. His philosophical style was very original and of great influence on those who witnessed his lectures. His main inspirers were Plato, Scheler and Bergson. Some of his publications are included in the literature list.

209 Both Henri Bergson and Edmund Husserl have explicitly recognized this aim in the beginning of our century. But their philosophies were very much involved in a struggle with the then prevalent positivism and scientism. This gave them a one-sidedness, which contributed to the philosophical 'schizophrenia' of this century, and the precise nature of which must still be clarified. My conjecture is that Bergson contents himself too easily

but it is by the measure in which this skill is mastered, that a philosopher is called great. This could be the reason why Aristotelianism has survived in the modern era. It is true that in the Aristotelian tradition conceptual and philosophical thinking have been hopelessly mixed up, and modern mathematism was a true liberation from this entanglement. But the primary principles made explicit in Aristotle's philosophy are so 'close to life,' that they express a truth about the physical world, which modern science has not yet been able to reconstruct properly. It was only by the ideological dogma that we cannot do anything more than *reconstruct* the world of experience, and that it is principally out of the range of our intellect to *understand* it, that Aristotle could be kept outside. The fierceness of the early moderns against him witnesses to his threatening presence there.

Only now are we in a position to consider the philosophical tradition critically. We need no longer be haunted by the shadows of the past, fearing that if we let Aristotle in, the authorities of church and state will follow, as the moderns did. We need not picture the mathematical sciences as an degenerate form of philosophy, nor do we have any longing for 'formalization' of philosophical methods. We clearly recognize mathematism in the tendency to judge content by structure, to construct and reconstruct without epistemological reflection on the origin of our constructive ideas, to presuppose ontological uniformity and reduce potentiality to fields of 'possible' beings, in fact thought of as actual, and last but not least in the tendency to over-systemize philosophy or make it look like a mathematical science. We need no longer reject metaphysics as a construction with the claim to be necessary. Nor do we need to reduce the origins of perspectives to merely factual circumstances, such as social networks or cultural prejudices. We can without shame look for what is essential in the nature of such perspectives, including the sociological and the culture-critical ones. We shall also be free to develop an undogmatic appreciation of tradition, classical as well as modern. There will be no need to defend one of them against the other. The synthesis has begun, and that is true post-modern philosophy.

---

with a dualism of living and lifeless matter and Husserl's notion of 'evidence' is identified too much with necessity. This makes it difficult to understand that we implicitly know necessities. Max Scheler tried to solve this problem with the help of his notion of 'the functionalization of the a priori.' But in my opinion he described the situation 'upside down': as if we first have clear understanding of necessary coherence which afterwards 'sinks back' into practical life and is so functionalized, instead of understanding functional knowledge as its original form, and its explicit form as a result of philosophical reflection.

INDEX OF PERSONS

Adorno	146
Aertsen	149, 172
Alembert	149
Aquinas	172
Aquino	152
Archimedes	12, 103
Aristotle	12, 19, 20, 23, 26, 37-40, 44, 56, 62, 70, 74, 110, 165, 167, 171, 178, 181
Aristotle's	48
Ayer	153
Barendse	172, 178
Berger	110
Berghuys	124
Bergson	29, 57, 73, 122, 129, 180
Berkeley	143
Bernays	98
Beth	36, 156
Boole	48
Borges	70
Boukema	88
Brouwer	37, 42, 97, 101, 173
Brouwer's	43, 51
Bruin	172
Buber	178
Campbell	99
Cantor	37, 79, 100-103, 168
Caroll	
Lewis	19
Cassirer	74
Cherniak	55
Church	100, 102, 168
Coolen	52
Curry	50, 100, 102
Curry's	51
Cusanus	15, 75
Demjanjuk	47
Derrida	20, 169
Descartes	12, 25, 37, 51, 63, 76, 106, 110, 113, 173
Dijksterhuis	14, 105
Dilworth	98, 99
Dingler	22, 40
Dreiser	96

Edman	89
Erasmus	155
Euridice	19
Falkenburg	41
Feuerbach	107
Fleischhacker	109, 122
Foster	16
Foucault	94
Frege	13, 21, 37, 48, 49, 60, 88, 100
Gaiser	20
Galenus	143
Galileo	11, 15, 92, 155
Galloys	139
Gehlen	53
Gies	86
Girard	94, 178
Gloy	149
Gödel	13, 37, 50, 78, 100, 102, 103
Hakfoort	96
Harré	99
Hegel	12, 13, 16, 20, 25, 37, 48, 53, 55, 57, 73, 83, 85, 86, 92, 104, 108, 111, 120, 122, 129, 152, 157, 159, 166, 172, 175, 177
Hegel,	83
Hegel's	9, 68, 88
Heidegger	13, 35, 36, 53, 146
Heraklite	179
Heraklitos	110
Herz	54
Hilbert	40, 100
Hollak	52, 53, 55, 73, 85, 105, 109, 172, 175
Hollak's	55
Hösle	145
Hoyer	69
Hume	48
Husserl	13, 19, 22, 40, 48, 71, 87, 95, 134, 143, 178, 180
Jourdain	79
Kant	26, 63, 106, 111, 129, 158, 167
Kesselring	157
Kiergkegaard	122
Kierkegaard	110
Klages	122
Koyré	112, 167
Kripke	30, 61
Kronecker	16
Kuhn	99

Lacan's	69
Lakatos	98
Laplace	165
Latour	94
Leibniz	37, 48, 108, 139, 173, 175
Lesniewski	48
Lévinas	169, 178, 179
Liebrucks	65
Liotard	169
Locke	47, 143
Lullus	48
Luther	155
Lynkeus	155
Mandelbrot	86
Marx	107, 110, 122, 175
McDermott	55
Miller	83
Monk	152
Mostowski	80
Munnik	172
Neuser	177
Newton	15, 34, 94
Nietsche	122
Occam	38, 63
Oldewelt	40, 129, 180
Orpheus	19
Ortega Y Gasset	53
Ostwald	96
Oughourlian	94
Parmenides	72
Peano	24
Peirce	159, 178
Penrose	37
Piaget	157
Pirsig	77
Plato	9, 12, 16, 19, 20, 33, 37, 38, 42, 61, 167, 169, 171
Plato's	59
Plessner	73, 129
Plotinos	37
Popper	40
Putnam	13
Rosenstock	178
Russel	98, 100
Russell	37, 50, 79, 103
Sapir	69, 70

Sartre	73, 110, 122, 178
Saussure	75, 92
Scheler	15, 25, 53, 122, 129
Schelling	109, 177
Schroeder	48
Scott	168
Shanker	24
Shelldrake	173
Simon	175
Simplicio	92
Socrates	166
Sophists	61
Speusippos	16, 37
Spinoza	108, 173, 175
Stenlund	11, 73, 100
Swift	72
Tarski	50, 100, 113
Taureck	142, 144
Theillard de Chardin	173
Thom	57
Thoreau	96
Tijmes	95
Tolkien	145
Toth	61
Uylenspiegel Tijl	61
Vàrdy	99
Veatch	26, 66
Velthoven	75, 172
Weyl	101
Whewell	99
Whorf	69, 152
Wilson	47
Wittgenstein	13, 19, 24, 25, 81, 83, 88, 94, 103, 152, 153, 159, 169
Wittgenstein's	51, 82, 83
Wolff	157
Xenocrates	16, 37

INDEX OF SUBJECTS

εποχη,	95
Λογος	
'logic' comes from	48
Absolute unity	
God as the	42
Abstinence	
in thought	40
Abstract	
elements	37
form of the understanding	52
technical concept is	54
Abstraction	
act of	40
mathematical	23, 27
Abstractionism	
compared with other positions	42
right about relationship to world of experience	43
Abstractionist	
compared to neo platonism	42
element of mathematical thinking	46
Academy	
mathematisation of the ideas in the later	37
Accessibility	
of facts	39
Accident	
and substance	26
Accumulation	
of capital	53
Actio in distans	24
Active cause	
of change	44
Actualization	
structuring as	29
Alice in Wonderland	
Red Queen	46
Analysis	
as a cognitive function	27
Analytical	
geometry	26
thought	20
Applicability	
of a model	47
of math...and platonism	38

of mathematical structures	40
of mathematics	41
Applications	
of mathematics	27
Arbitrariness	
of the rel. betw. sign and object	32
Aristotelian	
metaphysics	36
Aristotelianism	181
Arithmetic	
as a traditional discipline	25
Arranging	
individual elements	45
Art	
mathematics as an	40
Artificial intelligence	
all information technology is	54
Axiom of complete induction	100
Axiomatic	
formal systems	49
Axiomatisations	
for fields of mathematical objects	40
Axioms	
characterising structures	20
Beetle in the box	19
Beyond	
material reality	45
Body	
as the first instrument	53
Calculation	
and technology	51
Cantor-discontinuum	29
Catastrophe	
theory	57
Category	
mathematical category theory	32
Causa sui	168
Chain of instruments	
machine as a	53
Change	
as such	23
every material thing capable of	43
of rules	46
structural	24
Changeability	

as indeterminateness	44
Chaos theory	57
Chemical	
compound	20
Classical	
logic	46
mathematics	33
positions	43
terminology	43
Cogito sum	51
Combinations	
of elements of a model	47
Commands	
and information processing	55
Common citizen	
problems simple for the	38
Common sense	
world vs mathematical reconstruction	58
Complex	
numbers	32
Complicated	
objects	38
Computer	
as an instrument	53
design and use of	54
Computer science	33
Concept	
and realisation	53
as result of technical invention	52
ideal	22
role of	22
Concepts	
made explicit	46
Conceptualising	
as a cognitive function	27
Concrete case	
form and matter in a	44
Concrete things	
do not have independent principles	45
Connection	
immediate	25
Connectives	99
Consistency	176
Constructed	
strucutre can always be understood as	45

Constructing
    mathematical objects   28
Construction
    of math. objects   43
Constructions
    in mathematics and technology   51
    mental   42
    symbolic   27
Constructivism
    intuitionistic   24
Constructivism.
    Brouwer's   37
contiguity   24
Continuous   25
    trajectory   47
Continuum
    as a representation of change   47
    mathematical   29
Convenience
    of the use of a notation   31
Conventional
    representation   32
    symbols   31
Conventional symbols
    numerals as   31
Coordinate system   128
Correspondence-principle
    in technology   54
Counting
    and measuring   27
Creation
    of mathematical structures   40
Culprit   47
Culture
    Greek   25
Daily life
    full of implicit premisses   46
Data
    ordering of (Positivism)   38
Decimal
    notation   31
Deduction
    not the same as exactness   46
Deductive
    reasoning   46

190

Defender	47
Degenerated	
mathematical objectivity	42
Degree	
of structurability	56
Demarcation	98
Denotation	
as a kind of naming	30
Desire	94
Determinations	
are never complete	43
Deutera ousia	19
Dictionary	
lexicographic order of a	38
Discontinuum	
of Cantor	29
Discrete	25
Divisible	
apple	20
Division	
of the experienced world	26
Domain	
of mathematical objectivity	47
Drawing	
of figures	31
Dynamic	
character of math thinking	43
structure	57
Dynamics	
mathematical	57
Effects	
power over natural	41
Elementary	
geometry	31
Emanate	
plurality from God	42
Emanation.	
neo platonic perspective of	41
Emanatism	
right concerning dynamics of math. thinking	43
Emanative	
element of mathematical thinking	46
Emancipation	
of mathematics in the 19th cent.	33
Emotion	

neglect of	40
Empirical	
has many implicit premisses	46
individuals	47
investigation	47
level of knowledge	58
reasoning never exact	47
Empiricism	64
Empiricists	91
Entities	
virtual	21
Epistemological	
considerations	23
dilemma	43
Epistemological argumentation against metaphysics	149
Epitheoretical	
reasoning	50
Essence	
of material being	43
Ethical argument against metaphysics	151
Euclidean	
space	26, 29
Evolution	
of earthly life	57
Exact	
mathematical thinking is most	46
Exactness	
mathematical	46
not the same as deduction	46
Existence	
independent	24
Experience	154
sensory	23
world of	23
Experienced	
space	26
world	25
Experienced world	
division of the	26
Experiment.	
In science	54
Explicit	
construction of a technical concept	53
Exposition	
proof by	47

Extended	
in some space	29
Extension	
as represented by mathematical objects	29
as sufficient but not necessary for structurability	29
of the body	53
Extensions	
of existing mathematical structures	32
External	
parts	45
symbols	51
world	51
External identity	104
External reflection	159
External states	
corresponding to inner images	54
Facts	
and mathematical order	38
Fallibilism	116
Fallibilistic	148
objection against metaphysics	153
Fantasies	
math. obj. are not purely arbitrary	38
Fata morgana	
fata	43
Field	
of interest	47
Figures	
drawing of	31
Fingerprints	47
Finitism	
Wittgensteinian	24
Fixed point	104
Form	
as determination	44
of matter	41
Formal systems	
epitheoretical reasoning about	50
Formalisations	
syntactical or semantical	50
Forms of technology	52
Foundation	
of mathematics	51
Fregean	
logicism	37

Functionalistic argument against metaphysics	154
Fundamental	163
intuition of mathematics	40
Fundamental perspective	
characterising a type of technology	53
Futility argument against metaphysics	151
General premises	
reasoning from	46
Generalisation	40
Generating	
new structures in mathematics	32
Generation	
of math. objects	43
Generic systems	100
Geometrical body	20
Geometry	
analytical	26
as a traditional discipline	25
elementary	31
God	96
in emanative idealism	42
Gödel's	13
Greek	
culture	25
Group-theory	33
Handcraft-like	
technology	51
Handcraft-type	
technology	53
Hegelian dialectics	171
historical theory	154
Historical type of argument against metaphysics	148
Human intentions	
and information technology	55
Human practice	
and technology	51
Human soul	20
Idea	
technical	52
Ideal	
character of principles	46
concept	22
construction	46
entities	38
material	22

objects	38
of justice (example)	39
possibility	45
principles	46
properties	51
structurability	24
structure	22
structures	23
Ideal character	
of mathematical objectivity	47
Ideal determinability	
structurability is	45
Ideal meaning	
of principles	45
Ideal nature	
participation in	39
Ideal structures	
the semantical relating of	51
Ideales Sein	19
Idealisation	
mathematical abstraction as	45
Ideality	
mathematical	24
Ideas.	
mathematisation of	37
Identification	
of individuals	47
Identified	
individuals in a model are always	47
Identify	
the individuals	46
Identity	
individual	47
Ideology	154
Immaterial	
spirits	42
Immediate	
connection	25
representation	31
Implementation	
of a morphogenetic field	27
of a structure	22
Implicit premises	
in daily life and empirical science	46
Incompleteness	

theorems	50
Independent	
existence	24
Independent content	
of principles	45
Independent determinateness	
of the ideal	39
Indeterminateness	
matter as	43
ontological	43
Indifference	
between nature and structure	41
Indifferent	
matter and form as completely	45
Individual	
identity	47
mathematical	47
parts	26
Individuality	
of the sign	32
Individuals	
empirical	47
mathematical objects as	37
the same ?	46
Industrial	
technology	53
Information processing	
in technology	52
Information technology	
as self-reflection	54
Inner images	
of external states	54
Inner unity	173
mathematical objects do not have	37
Instrument	
as typical artefact of handcraft type technology	53
Intelligibility	169
of the world	39
Intelligible	
matter	23, 40
Interaction	
process	23
with nature	53
Interpretation	
of logical symbols	50

Interrelated	
individual parts	26
Intersubjectivity	177
Interval	
closed	23
Intuition	
mathematical	51
of the principle of structurability	51
purely spiritual	42
Intuitionistic	
constructivism	24
Intuitive	
in the limits of math. thinking	51
Invention	
and technology	51
as realisation of a concept	52
technical	40
Kinds	
natural	26
Know how	
characteristic of handcraft-type technology	53
Lambda calculus	168
Language	153
Language.	153
Laws of nature	
and technology	51
Levels	20
of knowledge	58
Lexicographic	
order (example)	38
Life	
as a possible exception to structurability	56
Light of understanding	
extinguishes in the face of life	57
Limit	
of a form of thinking	50
Limits	
of information technology	56
Lines	
Euclidean	26
Linguistic argument against metaphysics	152
Living	
organism	27
Living organisms	
as systems	58

Locality principle	24
Lock an key	
ancient form of information technology	53
Lock and key	
as paradigm of information technology	55
Logic	
classical	46
mathematical	46
mathematical thinking and	46
philosophical	48
What; Relating	26
Logical circuits	
design and use of	54
Logical connections	
representation of	50
Logical consequence	
of the axioms	46
Logical reasoning	
perfect applicability of	47
Logical symbols	
interpretation of	50
Logicism	
Fregean	37
Machine.	
idea of the	53
Making a person	
as defining AI	55
Manipulation	
syntactical	51
technical	22
Master Program	96
Material	
being	43
ideal	22
structure	23
thing	43
world	42
Mathematica	28
Mathematical	
abstraction	21, 23, 27
biology	57
exactness	46
form of technology	51
ideality	24
intuition	51

level of knowledge	58
logic	33, 46
objectivity	23, 24, 39
objects are too poor in content	37
order	38
structure	20
theories	49
thinking	28, 40, 51
thinking as exact	46
thought	47
Mathematical entities	19
Mathematical ideal	
matter tries to live up to it	42
Mathematical idealism	19
Mathematical logic	
and information technology	54
and mathematics	48
as self reflection	50
Mathematical objectivity	
as a projection screen	51
degenerated	42
domain of	47
Mathematical objects	20
Mathematical reconstruction	
of organic life	57
Mathematical structure	
principle of	43
Mathematical thinking	
and technology	52
self-reflection of	50
Mathematics	
and mathematical logic	48
and technology	51
classical and modern	33
proper	28
pure	27
the word	28
Mathematism	96, 157
and neo platonism	42
mathematism,	36
Matter	
form of	41
intelligible	23
tries to live up to mathematics	42
Matter.	

intelligible	40
Matter'.	
as indeterminateness	43
Meaning	19
Meaningful	
unities	28
Meaningless signs	99
Measurement	
and technology	51
attempt towards	22
Measuring	
and counting	27
Mechanistic	
approach to life	57
forms of physicalism	155
Mechanistic views	11, 14, 96
Mental	
constructions	42
Metamathematics	99
Metamathematics.	33
Metaphysical	
considerations	23
problem	47
Metaphysical dimension in language	152
Metaphysical nature	
of Aristotle's argument against Platonism	38
Metaphysics	99
Aristotelian	36
Pythagorean	36
Methodology	93
Mimesis	94
Model	
reasoning about a	47
Modernism	
in mathematics	33
More geometrico	175
Morphogenetic field	
implementation of a	27
Motion	
as represented by mathematical objects	29
of thoughts	39
Mountain	
mathematical thinking not like	40
mystical argument against metaphysics	151
Mythologies,	158

Name
- of a mathematical individual — 47

Natural
- kinds — 26

Natural language
- processing — 56

Natural numbers — 100
- the structure of — 30

Nature
- as it appears to sense perception — 37
- as it is in itself — 37
- of mathematical objects — 43

Necessary consequences
- by nature — 54
- in thought — 54

Neo-platonism
- mainstream of — 42

Non-Cantorian
- set theory — 23

Non-standard
- interpretations — 50

Notation
- drawing of figures as a form of — 31
- mathematical — 30
- systems — 31

Number
- notion of — 22

Numbers.
- complex — 32

Numerals
- as conventional symbols — 31

Objectification
- intuition and .. — 68
- of God's self-relatedness — 107
- of the technical idea — 53
- self-.. — 104

Objectifications
- of the subject in the world — 107

Objective
- in the sense of 'not arbitrary' — 28

Objectivity
- mathematical — 23, 24

Occam's razor — 38

Ontological
- indeterminateness — 43

Order
    of data (Positivism)   38
Organism
    living   27
Parallel-postulate
    validity of   23
Participates
    world in ideas (Plato)   38
Participation
    of the perceptible world   39
Particular
    form   44
    matter   44
Particular elements
    of our field of interest   46
Parts
    in math. objects   38
    individual   26
Passive potency
    matter as   44
Perceptible world
    assume anything outside the   38
    must have something to do with ideal entities   38
    participation of   39
Perfect applicability
    of logical reasoning   47
Permanence
    explained by matter   44
Permit for methodological chaos   156
Person   173, 178
Personality   104
Perspective   154, 178
    on mathematical objectivity   28
Perspectives
    three phil of math. as   43
Phantasy
    in mathematical thinking   45
Phenomenologically
    transcendent   47
Philosophical
    level of knowledge   58
    logic   48
Philosophical arrogance   166
Philosophical exactness   175
Philosophical tradition   181

Philosophy
    of mathematical abstraction and objectivity   43
    of mathematics   23
Philosophy of mathematics
    three types of   43
Physical
    interaction   23
    reality   25
    structure   24
Physicalism
    pre-modern forms of   155
Pico della Mirandola   16
Platonic
    element in mathematical thinking   46
Platonism   37
    right about the ideal nature of math. obj.   43
Platonist
    against positivist   38
Platonists   37
    in the philosophy of mathematics   37
Plurality
    emanation of   42
Plurality of principles   169
Poetry   153
    mathematics is not   38
Points
    Euclidean   26
Polynomial   32
Polynomial time   39
Positivistic
    argument against Platonism   38, 39
    objections to platonism   38
Postulation
    of mathematical rules   47
Potency
    structurability as a   29
    structurability is a   29
Power
    over natural effects   41
Power over nature
    and technology   55
Practical discipline
    logic as a   48
Presuppositions
    made explicit   46

Principle
- of math. structure — 43
- of structurability — 23, 26, 27, 40-42
- of technology — 52
- real — 25

Principle of non-contradiction — 157
Principle of principles — 166
Principle of structurability — 156
- intuition of — 51

Principles.
- form and matter as — 44

Process
- interaction- — 23

Processes
- power over natural — 41

Proof
- by exposition — 47

Properties
- without something having them — 21

Property
- and structure — 26

Pyramide — 92
Pyrrhonism — 155

Pythagorean
- metaphysics — 36
- positions — 24

Qualities
- mathematical objects do not have — 37

Quality
- and quantity — 26

Quantifiers — 99

Quantity
- and quality — 26
- Aristotle describes — 20

Quasi substances — 51
Quasi-substances — 173

Quasi-substantial
- structures — 39

Quicksand,
- we live on — 39

Re-structuring
- nature — 40

Real
- material world — 44
- principle — 25, 42

Realisation
    of the principle of structurability   23
Realisation.
    of a concept   53
Reality
    physical   25
Realphilosophie   166
Recombining
    in a different way   21
    the elements and laws of nature   40
Red Queen
    in Alice in Wonderland   46
Reflected
    the ideal in the real   38
Reflection   65
Reflection.   154
Reflective
    motion of thought   34
Regulative ideas   158
Reification   26
Relating logic   26
Relative opposition   172
Relatively opposed
    principles   44
Representation
    conventional   32
    immediate   31
    of logical connections   50
    symbolic   40
    systematic   32
Romantic   94
Rules
    of logic   46
Rules of the game
    may be changed   46
Sceptical
    aporia   46
Scepticism   155
Scientific
    technology   51, 53
Screen
    mathematical objectivity as a projection screen   51
Self-expression
    of a spiritual unity   42
Self-identification   104

Self-objectification	104
and self-awareness	111
in Hegel's philosophy	109
Self-reflection	65
and information technology	54
logic as	49
mathematical logic as	50
Self-reflection.	
of mathematical thinking	51
Semantical	
formalisations	50
relating of ideal structures	51
Semantics	100
in mathematical logic	49
Sensation	
neglect of	40
Sense-perception	
nature as it appears to	37
Sensory	
experience	23
Sensus communis	133
Separable	20
Separate existence	
of ideal objects	39
Series	
as a representation of change	47
Set-theory	33, 164
non-cantorian	23
Sign	
drawing as a	31
Signifiant	92
Signifié	92
Silent	
presuppositions	47
Simulation	
of thinking	54
Social structure	178
Sociological theory	154
Something	172
Space	
Euclidean	26, 29
experienced	26
topological	27
Speculative dialectics	157
Spirits	

immaterial	42
Spiritual soul	20
Structurability	96
as a potency	29
as ideal determinability	45
belonging to the essence	43
ideal	24
of the world	22
point of view of	22
principle	26
principle of	23, 27, 40
Structurability.	
indeterminateness as	44
Structurability'	
as a potency	29
Structural	
aspect of reality	40
change	24
Structural correspondence	
used in information technology	56
Structure	
abstract	22
and property	26
ideal	22
material	23
physical	24
Structure.	
determination becomes	45
Structures	
ideal	23
ready made	23
Structuring	26
as a cognitive function	27
as actualization	29
of a field of experience	27
of the world of experience	27
the world	22
Subjective	
idealism	42
Subjective side	
of facts	39
Substance	172
and accident	26
Substantial	
elements	26

parts and wholes equally substantial	38
subject matter	23
Substantialising	
as a cognitive function	27
Summa ius summa iniuria	39
Summing up	
of facts	38
Syllogism	
schemes (Aristotle's)	48
Symbol	
drawing as a	31
use of	49
Symbolic	
constructions	27
representation	40
Symbols	99
conventional	31
external	51
Syntactical	
formalisations	50
manipulation	51
Syntactical structure	99
Syntax	
in mathematical logic	49
Systematic	
representation	32
Tautology	
in mathematics	46
Technical	
idea	52
invention	40
manipulation	22
Technical idea	
objectification of	53
Technological	
aims	51
Technology	
handcraft-type	53
mathematical thinking fruitful for	51
Teddybear	104
Telephone directory	
lexicographic order of	38
Test case	
phenomenon of life as a	57
theological argument against metaphysics	151

Theorem	
mathematical	46
Theory	
of abstraction	39
Thinking	
mathematical	28, 51
Thinking bareheadedly	156
Time	
spatial image of	29
Topological	
space	27
specification of dimension	29
Topology	33
Totalitarization	178
Trajectory	
in a dynamic structure	57
Transcendent	
phenomenologically	47
Transitional object	104
Triangle	
Locke's	47
Truth functions	49
Truth-table	50
Type of technology	
characterised by two dimensions	53
Typical artefact	
characterising a type of technology	53
Ultimate content	
of principles	44
Ultimate individuals	
in a model	47
Ultimate rules	
in a model	47
Unchanging	
individuals in a model are	47
Understanding	
physical reality as such	57
the human spirit	57
Unities	
meaningful	28
Unity and plurality	172
Universal objectivity	
in technology	52
Upland plain	
mathematical thinking as an	40

Use
- of a notation   31
- of conventional symbols   31

Vicious circle principle   98

Virtual
- entities   21

Vis cogitativa   133

Vitalistical
- approach to life   57

What-logic   26

Wholes
- math. objects as   38

Wittgensteinian
- finitism   24

World
- experienced   25
- external   51
- of experience   23, 40, 42, 43

World of experience
- meaningful aspects of   51

LITERATURE

Adorno, Th.W., *Negative Dialektik*, Suhrkamp, Frankfurt a/M, 1966

Aertsen Jan, Denken van de Eenheid, *Tijdschrift voor Filosofie*, 52(1990) p.365ff

Apostle H.G., *Aristotle's Philosophy of Mathematics*, Chicago 1965.

Aquinas Thomas, *De Quator Oppositis*, (Opera Omn., Op.Min.I.,Parijs, 1949.

Aquinas Thomas, *De Veritate*, (Turijn\Rome 1964).

Aquinas Thomas, *In Boetium de Trinitate*, (Opusc.Omn.3, Parijs, 1927.)

Aquinas Thomas, *In duodecimos libros metaphysicorum Aristotelis expositio (In Metaph. Arist.)*, Marietti (Turijn\Rome 1964).

Aquinas Thomas, *Summa contra Gentiles*, (Turijn, 1938).

Aquinas Thomas, *Summa theologiae*, Latin text and English translation, London, Blackfriars, 1964-1970.

Aquinas Thomas, *Summa Theologica*, (B.A.C. Madrid 1951).

Aquinas Thomas, *Catena Aurea in quator Evangelia*, Taurini: Marietti, Roma 1925 (Sent.)

Aristotle, *The works of Aristotle*, Translated into English under the Editorship of W.D. Ross, Oxford, Clarendon Press, 1924.

Aristotle, *Aristotle in twenty three volumes*, The Loeb Classical Library, Harvard University Press, Cambridge Mass. 1969

Barendse B.A.M., Intersubjectief verkeer en lichamelijkheid, in: Barendse B.A.M. *Zich door het leven heendenken*, Kampen (Kok) 1982.

Barendse Chr., *"Zich door het leven heen denken"*, Keuze uit het werk van Barendse B.A.M, Kampen 1982.

Barendse Chr., De metafysische grondslagen der vrijheid, in Wilsvrijheid, verslag alg. verg. v.d. Ver. v. Thomistische Wijsbegeerte, Utrecht, 4,5 mei 1946.

Barendse Chr., Intersubjectief verkeer en lichamelijkheid; de bemiddeling van het lichaam, in *"Lichamelijkheid"*, Utrecht\Brussel, 1951.

Barendse Chr., O.P. Over de graden in het zijn, *Tijdschrift voor Philosophie (11) 1949* ,p.155-202.

Berger Herman, *Zo wijd als alle werkelijkheid*, Ambo, Baarn 1977.

Berghuys J.J.W., De zelfstandigheid der materie, in: *Substantie* , Utrecht (Spectrum) 1966.

Berghuys J.J.W., *Grondslagen der aanschouwelijke meetkunde*, Groningen, etc., Noordhoff, 1952.

Bergson H., *Matière et Mémoire*, Paris

Bergson H., *Essay sur les données immédiates de la conscience*, Paris 1981. (gr.kw)

Bergson H., *Evolution créatrice*, Paris 1959, (gr.kw)

Beth E.W., *Mathematical Thought*, Dordrecht, 1965.

Beth Evert, *De wijsbegeerte van de wiskunde van Parmenides tot Bolzano*, Antwerpen/Nijmegen, 1944.

Boukema H.P., *Tijdschrift voor Filosofie,* Familiegelijkenissen. Wittgenstein als criticus en erfgenaam van Frege.

Brouwer L.E.J., *Collected works I*, ed.A.Heyting, Amsterdam\Oxford\New York, 1975.

Brouwer L.E.J.,Consciousness, Philosophy, and Mathematics, *Proceedings of the Tenth International Conference of Philosophy*,Pos H.J., Beth E.W., Hollak J.H.A.(ed.), Amsterdam 1949.

Cassirer Ernst, *Substanzbegriff und Funktionsbegriff. Untersuchung über die Grundfragen der Erkenntniskritik.*, Darmstadt 1969 (Berlin 1910).

Cassirer, Ernst, *Substance and function and Einstein's theory of Relativity*, transl. from german by W.C.Swabey and M. Collins, New York, Dover Books, 1953

Cherniak E. and McDermott D., *An Introduction to Artificial Intelligence*, Addison&Wesley, NY, 1985.

Coolen T.M.T., *De machine voorbij,* Over het zelfbegrip van de mens in het tijdperk van de informatietechniek, Boom, Meppel, 1992.

Coolen T.M.T., Philosophical Anthropology and the Problem of Responsibility in Technology, *Technology and Responsibility* (Paul.T. Durbin (ed.), Reidel, Dordrecht 1987.

Curry H.B., & Hindley J.R., Seldin J.P., *Combinatory Logic II*, Amsterdam, 1972.
Curry H.B., Feys R. & Craig W., *Combinatory Logic I*, Amsterdam, 1968.

Curry H.B., *Outlines of a Formalist Philosophy of Mathematics*, Amsterdam, 1951.

Curry Haskell.B., *The Foundations of Mathematical Logic*, McGraw-Hill, New York, 1963.

Cusanus Nicolaus, *De coniecturis*, ed. I.Koch, C.Bormann, I.G. Senger., Hamburg 1972.

Derrida J., Le puits et la pyramide, *Marges de la philosophie*, Paris 1972.

Descartes R., *Oevres Philosophiques*, éd. Ferdinand Alquié, Garnier Frères, Paris 1963.

Descartes R., *Discours de la méthode*, éd. Ch. Adam & P. Tannéry, Paris, {1637},1897-1913.

Descartes René, Géométrie, in *Oeuvres* ed. Adam & Tannéry VI, Paris 1965.

Descartes René, *Règles utiles et claires pour la direction de l'esprit et la recherche de la vérité*, trad. par Jean-Luc Marion, Nijhoff, Den Haag, 1977

Descartes René, *The Philosophical Works of Descartes*, transl. Elisabeth S. Haldane and G.R.T. Ross, C.U.P. London, 1967.

Diehls H. & Kranz W., *Die Fragmente der Vorsokratiker*, Weidmann, Dublin, 1966/1967

Dijksterhuis E.J., *De mechanisering van het wereldbeeld*, Amsterdam, Meulenhoff, 1950,1975.

Dilworth C., Idealisation and the Abstractive-Theoretical Model of Scientific Explanation, *Poznan Studies in the Philosophy of the Sciences and the Humanities*, 1989.

Dilworth C., *Scientific Progress*, Dordrecht 1981 (1992).

Dilworth C.,The Fabric of Science, *Boston Studies in the Philosophy of Science* 1989.

Dilworth Craig, *Laws, Theories and the principles of science*, Uppsala 1990.

Dilworth Craig, Principles, Laws and Theories, *Boston Studies in the Philosophy of Science*;The Fabric of Science, 1989.

Dilworth Craig, Science and the Supervention of Experience, in *The Role of Experience in Science*, Proceedings of the 1986 Conference of the Académie Internationale de Philosophie des Sciences, held at the University of Heidelberg, Berlin/New York 1988.

Dilworth Craig, *The Metaphysics of Science*, Kluwer, Dordrecht (publication expected) 1995

Dingler Hugo, *Das Experiment: sein Wesen und seine Geschichte*, Reinhardt, München 1928.

Dreiser Theodore, *The Living Thoughts of Thoreau*, Fawcet Libr NY, 1958.

Dumouchel Paul & Dupuy Jean Pierre, *L'Enfer des choses*, Paris 1979.

Edman Erwin, *The Philosopher's Quest*, Viking Press New York, 1947

Encyclopedic Dictionary of Mathematics MIT press 1980.

Falkenburg Brigitte, *Die Form der Materie*, Frankfurt a/M 1987.

Fleischhacker, L.E., Is Russell's Vicious Circle Principle false or meaningless?, *Dialectica* 33(1979), pp.23-30

Fleischhacker, L.E., &J.Kuper, Deontic Logic and the Axiom of Necessity: the Consequences of a Misinterpretation, *Journal of Value Inquiry* 16(1982) pp.67-74

Fleischhacker, L.E., *Over de Grenzen van de Kwantiteit*, Dissertatie, Amsterdam 1982 (251 pag.)

Fleischhacker, L.E., Moet fysica mathematisch zijn?, *Nederlands Tijdschrift voor Natuurkunde*, B49(13) 1983, pp.83-85

Fleischhacker, L.E., Over de verslaving in dienst van het gesplitste atoom, in Th.v.Velthoven (ed.) *Zin en Zijn*, Baarn 1983 pp. 89-100.

Fleischhacker, L.E., Wiskunde, taal en muziek, *bundel 4e Nederlandse filosofiedag, Groningen* 1984.

Fleischhacker, L.E., Het Contradictore Octaaf, *Kennis en Methode*, 1984/2 pp.106-124

Fleischhacker, L.E., De actualiteit van Hegels leer van de zedelijkheid, *Praesidium Libertatis*, bundel Filosofiedag Leiden 1985, pp.207-213

Fleischhacker, L.E., Hegels theorie van de oorlog in het licht van de kernbewapening, *Werken met wijsbegeerte* (Red. G.M.Huussen, H.E.S.Woldring), Delft 1986, pp.269-277

Fleischhacker, L.E., Quantität, Mathematik und Naturphilosophie, in *Hegel und die Naturwissenschaften* (Red. M.Petry), Stuttgart/Bad Cannstatt 1987, pp. 183-203

Fleischhacker, L.E., Het Wiskundig Teken, in *Reflexiviteit en Metafysica* (Red. L.E.Fleischhacker), Delft 1987, pp. 20-38

Fleischhacker, L.E., Creativiteit in arbeid en automatie, *Algemeen Nederlands Tijdschrift voor Wijsbegeerte*, 80(1988), pp. 272-288

Fleischhacker, L.E., Arbeid en (kunstmatige) intelligentie, in *Arbeid adelt niet* (Red. P.Tijmes), Kampen 1989, pp.135-154.

Fleischhacker, L.E., Gibt es etwas ausser der Äusserlichkeit, in *Hegel Jahrbuch* 1990, pp. 35-41

Fleischhacker, L.E., Zelfbeschikking en Wereldgeschiedenis, *Stoicheia*, 5(1990) 3, pp. 63-73.

Fleischhacker, L.E., Hegels Entwicklung der logischen Grundprinzipien in der Wissenschaft der Logik, in: *Hegels Transformation der Metaphysik*, Dinter Ver. Köln 1991, pp. 83-97.

Fleischhacker, L.E., Review of *Language and Philosophical problems* by Sören Stenlund, *EPISTEMOLOGIA*, 1992.

Fleischhacker, L.E., Mathematical Abstraction, Idealisation and Intelligibility in Science, in C.Dilworth (ed.), *Intelligibility in Science*, Poznan Studies in the philosophy of the sciences and the humanities, Amsterdam/Atlanta, GA 1992.

Fleischhacker, L.E., Het mathematisch ideaal, *De Uil van Minerva*, Gent 1993

Fleischhacker, L.E., The mathematisation of life, *Graduate Faculty Philosophy Journal of the New School for Social Research*, New York, 1993

Fleischhacker, L.E., Mathematics and experimental science in Hegels Philosophy of Nature, in M.J. Petry (ed.), *Hegel and Newtonianism*, Kluwer, Dordrecht, 1993, pp. 209-225

Foster, Michael N., *Hegel and Scepticism*, Harvard 1989

Gaiser Konrad, *Platons ungeschriebene Lehre*, Stuttgart 1963(1968).

Galilei G., *Dialogo sopra i due massimi sitemi del mondo*, Edizione nazionale VII, German translation by E. Strauss.

Girard, René, *Mensonge romantique et verité romanesque*, Grasset, Paris 1961

Girard R., *Des choses cachées depuis la fondation du monde*, Paris 1978.

Girard R., *La violence et le sacré*, Gasset Paris 1972.

Gloy K., *Einheit und Mannigfaltigkeit*, Berlin\New York 1981.

Hakfoort C.J.M., Science defied: Wilhelm Ostwald's energeticist world-view and the history of scientism., *Annals of Science*, 49(1992), pp. 525-544.

Hegel G.W.F, *Enyklopädie der Philosophischen Wissenschaften III*, Suhrkamp, Bnd.10, § 552. (wwc)

Hegel G.W.F, *Jenaer Realphilosophie* (ed. Felix Meiner, Hamburg 1931(1969).

Hegel G.W.F, *Phaenomenologie des Geistes*, ed. Hoffmeister, Felix Meiner, Hamburg 1952 p.52; Miller A.V. (Transl.), *Hegel's Phenomenology of Spirit*, Oxford 1977.

Hegel G.W.F., *De Orbis Planetarum, Philosophische Erörterung über die Planetenbahnen*, vert.Wolfgang Neuser, Weinheim 1986.

Hegel G.W.F., *Enzyklopädie der philosophischen Wissenschaften im Grundrisse, I-III, Werke in zwanzig Bände, 8-10*, Theorie Werkausgabe, Redaktion E. Moldenhauer & K.M. Michel, Frankfurt am Main, Suhrkamp, {1830}, 1970.

Hegel G.W.F., *Enzyklopädie der philosophischen Wissenschaften*, Suhrkamp, Werke in 20 Bände.

Hegel G.W.F., *Gesammelte Werke*, Deutscher Forschungsgem./Rh.Westf.Akad, Hamburg 1966-1976.

Hegel G.W.F., *Grundlinien der Philosophie des Rechts*,Suhrkamp Bnd.7, § 324. (wwc)

Hegel G.W.F., *Jenaer Realphilosophie*, Hrsg. J. Hoffmeister, Hamburg 1969.

Hegel G.W.F., *La théorie de la mesure*, trad. et comm. par A.Doz, Paris 1970.

Hegel G.W.F., *Naturphilosophie, Band I, Die Vorlesungen von 1819/20*, Hrsg. und Komm. M. Gies, Napoli 1980;

Hegel G.W.F., *Sämtliche Werke, Jubiläumausgabe in 20 Bänden*, Stuttgart 1964.

Hegel G.W.F., *Werke in zwanzig Bänden*, Suhrkamp, Frankfurt a.M. 1969-1971.

Hegel G.W.F., *Wissenschaft der Logik*, ed Suhrkamp: Werke in zwanzig Bänden 5/6, Frankfurt am Main 1969.

Heidegger, M., *Die Technik und die Kehre*, Pfullingen, Neske, 1962,1978.

Heidegger Martin, *Die Frage nach dem Ding*, Tübingen 1975.

Herz Heinrich Rudolf, *Die Prinzipien der Mechanik in neuem Zusammenhang dargestellt*, Leipzig, 1894.

Hollak J.H.A., Betrachtungen über das Wesen der heutigen Technik, Kerygma und Mythos VI, Band III, *Theologische Forschung 44*, 1968, Hamburg, Evangelischer Verlag, pp. 50-73, Vertaling van Italiaans artikel, zie ook onder Hollak.

Hollak J.H.A., Considerazioni sulla natura della tecnica odierna, l'uomo e la cibernetica nel quadro delle filosofia sociologica, *Tecnica e casistica*, Archivo di filosofia, 1/2, Padova, 1964 pp. 121-146, discussie pp. 147-152.

Hollak J.H.A., De Godsidee in de moderne metaphysica, *Vox Theologica* 36, 2, pp.62-74.(1966).

Hollak J.H.A., De structuur van Hegels wijsbegeerte (Dissertatie), *Tijdschrift voor Philosophie 24*, 1962, pp 351-403 en 524-614.

Hollak J.H.A., De wording van de menselijke geest, *Themata rond Evolutie*, Annalen van het Thijmgenootschap 51, Utrecht en Antwerpen, 1963, pp. 24-40.

Hollak J.H.A., Hegel, Marx en de cybernetica, *Tijdschrift voor Philosophie 25*, 1963, pp. 279-294.

Hollak J.H.A., Technik und Dialektik, *Civilisation, technique et Humanisme*, Paris, 1968, pp. 177-188, Colloque de l'Académie internationale de philosophie des sciences, Lausanne, 1965.

Hollak J.H.A., *Van causa sui tot automatie*, oratie Universiteit v. A'dam, Hilversum 1966. pp.47. ,

Hollak J.H.A., Vom materialistischen Dialektik bis zum dialektischen Materialismus, *Tijdschrift voor Philosophie 21*, 1959, pp. 518-528. (ook in:*Atti del 12 congresso internationale di filosofia 12*, 1961, pp. 233-241.

Hollak J.H.A., Von der Causa Sui Idee zur Automation, *Akten des XIV. internationalen Kongresses für Philosophie* II, Wien 1968, pp. 46-52.

Hollak J.H.A., Wijsgerige reflecties over de scheppingsidee; St. Thomas, Hegel en de Grieken, *De Eindige Mens*, essays over de grenzen van het menselijk bestaan, Bilthoven 1975, pp. 89-103 [Annalen van het Thijmgenootschap 63]

Hösle V., *Hegels System. Der Idealismus der Subjektivität und das Problem der Intersubjektivität*, 2 Bde., Hamburg 1987

Hoyer Harry, *Language in Culture*, Chicago 1954.

Husserl Edmund, *Die Krisis der Europäischen Wissenschaften und die tranzendentale Phänomenologie*, Den Haag, 1962.

Husserl Edmund, *Logische Untersuchungen*, Nijhoff, den Haag 1950 (Kluwer, Dordrecht 1987).

Husserl Edmund, *Ideeen zu einer reinen Phänomenologie und phänomenologischer Philosophie*, Nijhoff, Den Haag 1950-1952

Kesselring Thomas, *Die Produktivität der Antinomie, Hegel's Dialektik im lichte der genetischen Erkenntnistheorie und der formalen Logik*, Suhrkamp, Frankfurt a/M, 1984, ISBN 3-518-57653-4

Kesselring Thomas, *Entwicklung und Widerspruch, ein Vergleich zwischen Piagets genetischer Erkenntnistheorie und Hegels Dialektik*, Suhrkamp, Frankfurt a/M, 1981.

Koyré A., *Epiménide le menteur*, (Série:Actualités scientifiques et Industrielles, 1021), Paris s.d.

Kripke Saul, Naming and Necessity, in: *Semantics of Natural Language*, ed. D.Davidson & G. Harman, Reidel, Dordrecht, 1972.

Lakatos Imre, *Proofs and Refutations*,(Cambridge 1976).

Leibniz, G.W., *Discourse on Metaphysics*, Section XII, translation Lucas P.G. and Grint Leslie Manchester 1953(1968).

Leibniz G.W., *Die philosophischen Schriften*, hrsg. von C.I. Gerhardt, Hildesheim 1965.

Liebrucks Bruno, *Sprache und Bewußtsein*, Frankfurt a/M, 1964-1969

Monod J., *Le hazard et la nécessité*.Paris 1970.

Neuser Wolfgang, Einfluß der Schellingschen Naturphilosophie auf die Systembildung bei Hegel: Selbstorganisation versus rekursive Logik, in:*Die*

Naturphilosophie des deutschen Idealismus, Hrsg. Karen Gloy und Paul Burger, Frommann-Holzbeeg, Stuttgart 1993

Newton Isaac, *Opticks, Query 23/31 (1730 ed.)*, Dover reprint.

Oldewelt H.M.J., *Leven en Spreken*, Wereldvenster, Bussum, 1973

Oldewelt H.M.J., *Literaire reacties op de wereld van heden*, Servire, Hilversum, 1953

Oldewelt H.M.J., *Mensch en Wijsgeer*, Noord-Holland uitg., Amsterdam 1946

Oldewelt H.M.J., *Over ouders en kinderen schijnt dezelfde zon*, De Torenlaan, Assen 1959

Oldewelt H.M.J., *De plaats van de mens in de totaliteit van het leven*, Noord-Holland uitg., Amsterdam 1945

Oldewelt H.M.J., *Plato*, Kruseman, Den Haag 1931

Oldewelt H.M.J., *Proeve ener introspectieve Plato studie*, Paris, Amsterdam 1927

Oldewelt H.M.J., *Terugblik en uitzicht*, (Herdenkingsbundel) Kruseman, Den Haag 1967

Oughourlian G.M., *Un mime nommé désir*, Grasset, Paris 1982.

Petry, M.J. & Horstmann R.P. (Hrsg.), *Hegels Philosophie der Natur*, Stuttgart, 1986.

Petry M.J. (ed.), *Hegel und die Naturwissenschaften*, Stuttgart 1987.

Petry M.J., Hegels kritiek op Newton, *Wijsgerig Perspectief* 22.4(1981/1982)p103-112.

Petry M.J. (ed.), *Hegel and Newtonianism*, Kluwer, Dordrecht, 1993

Plessner Helmuth, *Philosophische Antropologie*, Fischer, Frankfurt a/M, 1970

Plessner Helmuth, *Conditio Humana*, Neske, Pfullingen, 1964; Suhrkamp, Frankfurt a/M 1983

Russel Bertrand, *An Enquiry into Meaning and Truth*, Pelican Edition. 1962

Sapir Edward, Conceptual categories in primitive languages, *Science*, **74**(1931).

Scheler Max, *Die Wissensformen und die Gesellschaft*, Bern 1969.

Shanker S.G., Wittgensteins Remarks on the Significance of Gödel's Theorem, in *Gödel's Theorem in Focus*, London: Croom Helm, 1988.

Sheldrake Rupert, *A New Science of Life*, Paladin, London 1983/1984

Sheldrake Rupert, *The Presence of the Past*, Collins Bodmin(G.B.) 1988

Stenlund Sören, *Language and Philosophical Problems*, Routledge, London 1990.

Taureck B., *Das Schicksal der philosophischen Konstruktion*, Wien 1975.

Taureck B., *Mathematische und tranzendentalen Identität*, Wien 1973.

Thom René, *Mathematical Models of Morphogenesis*, Ellis Horwood, Chichester 1983.

Tijmes P., The genius of the master lies in the limitation, in: *Journal of the American Academy of Religion* 1994

Tijmes P.& Luijf Reginald, Modern Immaterialism, in: *Research in Philosophy and Technology*, Volume 12(1992) p. 271-284.

Toth I., *Die nicht-euklidische Geometrie in de Phaenomenologie des Geistes*, Frankfurt/Main 1972.

Toth I., Non-Euclidean geometry before Euclid, *Scientific American,* November 1969.

Toth I., Spekulationen über die Möglichkeit eines nichteuklidischen Raumes vor Einstein, *Einstein Symposion*, Berlin 1979.

Vàrdy P., Some remarks on the Relatioship between Russell's Vicious-circle-principle and Russell's Paradox, *Dialectica 33*,(1979).

Vàrdy P.,*Zur Dialektik der Metamathematik*, colloquium over de Hegelse natuurphilosophie te Tübingen, najaar 1983.

Vàrdy P., Zur dialektik der Metamathematik, in: Petry M. (ed.), *Hegel und die Naturwissenschaften*, Frommann-Holzboog, Stuttgart 1987.

Veatch H.B., *Aristotle: a Contemporary Appreciation*, Bloomington & London, Indiana University Press, 1974.

Veatch H.B., *Realism and Nominalism Revisited*, Milwaukee 1954.

Veatch Henry B., *Two Logics*, Northwestern University Press, Evanston 1969.

Velthoven Th. van, *De intersubjectiviteit van het zijn*, keuze uit zijn werk met een inleiding van J.Aertsen, Kampen, Kok Agora, 1988.

Velthoven Th. van, *Gottesschau und menslichen Kreativität*, Leiden 1977 (diss.).

Velthoven Th. van, *Ontvangen als intersubjectieve act*, oratie Univ.v.Amsterdam 1980.

Wittgenstein Ludwig., *Philosophische Untersuchungen, Schriften I*, Frankfurt 1963.

Wittgenstein Ludwig, *Remarks on the Foundations of Mathematics/Bemerkungen über die Grundlagen der Mathematik*, ed. by G.H. von Wright & G.E.M. Anscombe, Basil & Blackwell, Oxford 1964.

Wittgenstein Ludwig, *Bemerkungen über die Grundlagen der Mathematik*, Hrsg. G.E.M. Anscomb, R. Rhees & G.H. von Wright, Suhrkamp, Frankfurt a/M 1974.

Wolff M., *Der Begriff des Widerspruchs, eine Studie zur Dialektik Kants und Hegels*, Meisenheim 1981.

Jarmo Pulkkinen

The Threat of Logical Mathematism
A Study on the Critique of Mathematical Logic in Germany at the Turn of the 20th Century

Frankfurt/M., Berlin, Bern, New York, Paris, Wien, 1994. 187 pp.
Scandinavian University Studies in the Humanities and Social Sciences
Edited by Hartmut Schröder. Vol. 7
ISBN 3-631-47409-1　pb. DM 64.--*

The present survey of the critique of mathematical logic in Germany at the turn of the 20th century attempts to answer several interesting questions: How did the contemporary German philosophers see the role and significance of logic? What kind of relationships did they claim to exist between logic, mathematics, linguistics and psychology? Pulkkinen starts by giving a historical survey of the development of German logic 1830-1920 as it appears against the background of German academic philosophy. Next he studies the interrelationships between logic and psychology, logic and linguistics, and logic and mathematics. After this the author presents the general features of the reception of mathematical logic in Germany between 1880 and 1920. This is followed by a more detailed account of the arguments of three individual critics: Fritz Mauthner, Heinrich Rickert, and Theodor Ziehen.

Contents: Survey of the reception of mathematical logic in Germany at the turn of the 20th century - Description of the development of German logic between 1830 and 1920 - Interrelationships between logic, mathematics, linguistics and psychology in the same period

Peter Lang ░░░ Europäischer Verlag der Wissenschaften
Frankfurt a.M. • Berlin • Bern • New York • Paris • Wien
Auslieferung: Verlag Peter Lang AG, Jupiterstr. 15, CH-3000 Bern 15
Telefon (004131) 9411122, Telefax (004131) 9411131

- Preisänderungen vorbehalten - *inklusive Mehrwertsteuer

Milton Keynes UK
Ingram Content Group UK Ltd.
UKHW021835111023
430426UK00009B/98